易象·六道佛 200cm×150cm

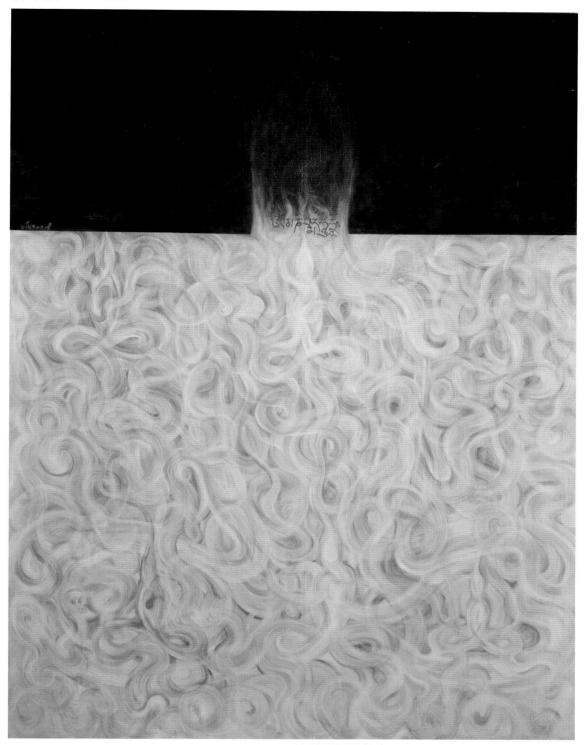

U0351080

易者，象也……

易象

WINDOO
温 度 职业艺术家 客家人
深圳梧桐艺术小镇可以艺术街
Tel:13682530717
QQ:40987733

Subscription:
Artpower Creative Space is a bimonthly magazine, which is published on 5th every other month. The price per issue is 18USD.

订阅邮购启事：
《ACS 创意空间》为双月刊，逢单月 5 日出版，每期定价 68 元，全年定价 388 元（增刊未含在内）。

05 2013/12 Bimonthly 双月刊
www.artpower.com.cn

Sponsor/ Dalian University of Technology (The University Press)
Artpower International Publishing Co., Ltd.

Consultants (In no particular order)
Nico van der Meulen (South Africa)
Kevin Kennon (America)
Recardo Bofill (Spain)

President & Editor in Chief/ Jin Yingwei
Editorial Director/ Yuan Bin
Planner/ Lu Jican
Brand Operation Director/ Meng Fengjun
Sales Director/ Deng Zhenggao
Overseas Sales/ Zhang Hong
Editor in Charge/ Qiu Meiqian
Proofreader in Charge/ Wang Dandan
Chief Material Collector/ Wang Yu
Overseas Editor/ Song Jia, Li Aihong
Copy Editor/ Xia Jiajia
Translator/ Ding Xiaojuan, Mo Tingli
Coordinator/ Ding Xiaojuan
Art Editor/ Chen Ting Wang Anlei Zheng Qin

Publication & Distribution/ Dalian University of Technology Press
Shenzhen Huangtai Printing Co., Ltd.

Publishing Date/ December, 2013
Price per Issue/ RMB 68.00 / HKD 98.00 / USD 18.00
ISBN 978-7-89437-083-9
Address/ Room 1102, Section B, Sic-Tech Building, No.80 Software Park Road, High-tech Industrial Zone, Dalian, China, Dalian University of Technology Press
Zip Code/ 116023

Editorial Department Tel/ +86 411 84709246 84709043
Editorial Department Fax/ +86 411 84709246
Advertising Department Tel/ +86 411 84709246
Sales Department Tel/ +86 411 84708842
Overseas Sales Department Tel/ +86 411 84709043
Delivery Department Tel/ +86 411 84703636

Shenzhen Editorial Department
Address/ G009, Floor 7th , Building 807-808, Yimao Centre, Meiyuan Road, Luohu District, Shenzhen, China
Advertising Department Tel/ + 86 755 82913355
Sales Department Tel/ + 86 755 82020029

主办单位 / 大连理工大学（出版社）
协办单位 / 深圳市艺力文化发展有限公司

顾问（排名不分先后）
尼科·范德·麦伦（南非）
凯文·凯诺（美国）
里卡多·波菲（西班牙）

社长兼主编 / 金英伟
编辑部主任 / 袁斌
策划人 / 卢积灿
品牌运营总监 / 孟逢君
发行总监 / 邓正高
海外发行 / 张泓
责任编辑 / 裘美倩
责任校对 / 王丹丹
采稿部经理 / 王宇
海外编辑 / 宋佳 李爱红
文字编辑 / 夏佳佳
翻译 / 丁小娟 莫婷丽
流程编辑 / 丁小娟
美术编辑 / 陈婷 王安磊 郑勤

出版发行 / 大连理工大学出版社
印刷 / 深圳市皇泰印刷有限公司

出版日期 / 2013 年 12 月
定价 / 68.00 元人民币 / 98.00 港币 / 18.00 美元
ISBN 978-7-8-89437-083-9
地址 / 辽宁省大连市高新园区软件园路 80 号
理工科技园 B 座 1102 室 大连理工大学出版社

邮编 / 116023

编辑部电话 / +86 411 84709246 84709043
编辑部传真 / +86 411 84709246
广告部电话 / +86 411 84709246
发行部电话 / +86 411 84708842
海外发行部电话 / +86 411 84709043
邮购部电话 / +86 411 84703636

深圳编辑部
地址 / 深圳市罗湖区梅园路 807-808 栋艺茂中心 7 楼 G009
邮编 / 518040
广告部电话 / + 86 755 82913355
发行部电话 / + 86 755 82020029

Statement:
1. The article published in this magazine only represents the author's opinion, rather than the opinions from the Editorial Board or Editorial Department. Any academic criticism or discussion for the content published from the reader is welcomed.
2. The magazine reserves the exclusive publishing right of the article and image in the forms of Chinese and English versions, digital and web editions. Without permission, anyone shall not, for the purpose of profit, copy, reprint, extract and compile, adapt, translate, note, collate and edit the article and image.
3. The magazine encourages the contributions from design agencies and the personal. The magazine reserves a right for editing. Any particular requirement shall be filed beforehand. Please do not duplicate submit.
4. The author is responsible for his submission. Quotation and image should credit the source. For those article and image which infringe the other's copyright or other rights, the magazine is not jointly and severally liable.

本刊声明
1. 本刊所刊载的文章仅代表作者的观点，并不完全代表编委及编辑部观点，欢迎读者对刊载内容展开学术批评和讨论；
2. 本刊保留所有刊载文章及图片的中英文、电子、网络版的专有出版权，未经许可，任何人不得以营利为目的复制、转载、摘编、改编、翻译、注释、整理、编辑等，本刊保留对侵权者采取法律行动的权利；
3. 本刊欢迎各设计单位及个人踊跃投稿，本刊对来稿保留修改权，有特殊要求者请事先声明，请勿一稿多投；
4. 作者文责自负，文中所引文献、图片要有出处，对于侵犯他人版权或其他权利的文稿、图片，本刊不承担连带责任。

无相禅意
是对生命的
一种思考
一丝明悟
不矫揉造作
工于自然
不刻意追捧
志存挚诚
不浮于嚣躁
还原本质
随缘而生
无有所住
至真无碍

活│生│意│禅

www.zend.hk

TM: 13905983590

CONTENTS 目录

05
总第 05 期
2013/12

SPACE CONCEPT COLUMN CROSS BOARDS WIDE ANGLE IDEAS
空间　概念　专栏　跨界　广角　创意

SPACE 空间

- **012　KPMG'S** DANISH HQ
  KPMG 丹麦办公室
- **016　FROYO YOGURTERIA** IN VOLOS, BRANDING DESIGN APPLICATION
  沃洛斯市 Froyo Yogurteria 店面品牌设计
- **020　MUMAC** MUSEUM OF COFFEE MACHINE
  MUMAC 咖啡机博物馆
- **028　BETTER PLACE**
  百特普雷斯
- **036　WASHINGTON,** D.C.
  华盛顿特区
- **040　PEDESTRIAN** BRIDGE
  人行天桥
- **044　SHOFFICE**
  棚屋办公室
- **048　TONY'S** FARM
  托尼的农场
- **052　SHREWSBURY** INTERNATIONAL SCHOOL
  曼谷什鲁斯伯里国际学校
- **056　VILLA** SSK
  SSK 别墅
- **064　DIGIT — DIESEL HOME** COLLECTION INSTALLATION
  数字——DIESEL 家居展示

CONCEPT 概念

- **074　AURORA BOREALIS ARCTIC OBSERVATORY —** THE WINGS OF THE DAWN GODDESS
  北极光天文台——曙光女神的翅膀
- **078　PHOENIX** OBSERVATION TOWER
  凤凰城观景塔
- **082　MARINE** RESEARCH COMPLEX
  海洋研究中心综合体
- **086　WESTBAHNHOF** U-BAHN
  WESTBAHNHOF 地铁站

20TH ASIA-PACIFIC INTERIOR DESIGN AWARDS

第二十届亚太区室内设计大奖入围及获奖作品集

ISBN: 978-7-5623-4029-4   Size: 285X368 mm   Pages: 464

This book includes the finalists and award-winning projects of the 20TH ASIA-PACIFIC INTERIOR DESIGN AWARDS. They fall into chapters, including gold medal, food space, hotel space, leisure & entertainment space, living space, sample space, installation & exhibition space, public space, work space, shopping space and students' projects. This collection will show the readers the wisdom and design ideas of the designers in Asia-pacific region and unveil the design lightspots which belong to the Asia-pacific region.

本书收录了第 20 届亚太区室内设计大奖优秀的入围及获奖作品，分为以下章节：金奖、用餐空间、酒店空间、休闲 / 娱乐空间、居住空间、样板房空间、设施 / 展览空间、公共空间、办公空间、购物空间和学生作品。这本优秀的作品选将为读者呈现亚太地区设计师们的睿智和设计心得，展现属于亚太地区设计的各种亮点。

联系人：卢积灿　　　　　　联系人：马孝山
订阅电话：**135 1085 8202**　订阅电话：**139 2381 7653**

artpower@artpower.com.cn

Inspiration from Nature and Man
Designer | Christopher David White
来自于大自然和人类的灵感

CONTENTS 目录

05
总第 05 期
2013/12

SPACE CONCEPT COLUMN CROSS BOARDS WIDE ANGLE IDEAS
空 间　概 念　专 栏　　跨 界　　广 角　　创 意

094 STORIES BEHIND PASSIONS – Robert Majkut
背后的故事·热火设计——罗伯特·迈酷

096 ING Corporate Banking
ING 企业银行

100 Kronverk Cinema
Kronverk 电影院

103 Multikino Złote Tarasy
Multikino Złote Tarasy 影剧院

106 Orange Cinemas
橙电影院

112 Whaletone
鲸鱼钢琴

116 Thread / Machine Embroidery Articles
线绣摆件

118 Nexuscouch
尼克瑟斯躺椅

121 Inspiration from Nature and Man
来自于大自然和人类的灵感

122 Sculpture
雕塑

Title of work | Banksia, Dynamic
Material | High-Fired, (1240 centigrade), unglazed porcelain
Technique | Hand-built, Multiple firings

作品名称：动态山龙眼
材料：高温焙烧（1240度）素瓷
工艺：手工制作，多重烧制

COLLECTION OF THE 20TH ASIA-PACIFIC INTERIOR DESIGN AWARDS

第二十届亚太区室内设计大奖参赛作品选

ISBN: 978-7-5623-4023-2   Size: 245 X 290 mm

LIVING SPACE
280 pages

FOOD SPACE
224 pages

WORK SPACE
216 pages

SHOPPING SPACE+PUBLIC SPACE+HOTEL SPACE
216 pages

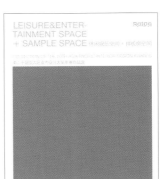

LEISURE&ENTERTAINMENT SPACE+SAMPLE SPACE
272 pages

INSTALLATION EXHIBITION
208 pages

联系人：卢积灿
订阅电话：**135 1085 8202**

联系人：马孝山
订阅电话：**139 2381 7653**

artpower@artpower.com.cn

CONTENTS 目录

05
总第 05 期
2013/12

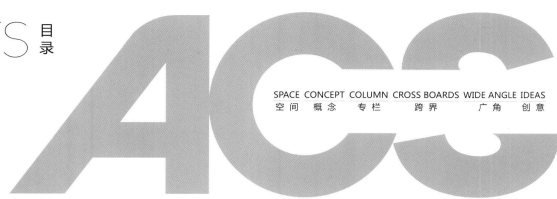

SPACE　CONCEPT　COLUMN　CROSS BOARDS　WIDE ANGLE　IDEAS
空间　概念　专栏　跨界　广角　创意

126	**VOLKSWAGEN** WIRE	大众电线车型
130	**HOMAGE** TO THE LOST SPACES	致敬基督城的失落空间
132	**LANDSCAPE** ABBREVIATED	景观缩放
134	**PTASZARNIA**	筑巢
136	**UBOJNIA**	屠宰场

140	**PIANTALÀ**	清新篱笆
143	**INFINITY** BENCH	"无限"长凳
144	**NOTHING** IS QUITE AS IT SEEMS	事事非所见
146	**THE** CANDELIER	小熊吊灯
148	**THE HOTWHEELS** WALL ART	"赛车"墙面艺术
150	**THE** ICE LAMP	冰灯
152	**PENDANT** CHANDELIER	垂饰吊灯
154	**36H** AND 56H	36 小时 /56 小时
156	**PONTE**	桥式桌
158	**"XII"** COLLECTION—MASARNIA	"XII"系列——MASARNIA
160	**WATERTOWER**	水塔
162	**LEVITY** CHAISE-LOUNGE	飘浮轻便躺椅

VOLKSWAGEN WIRE
Designer | Karen Oetling Corvera, Juan Pablo Ramos Valadez
大众电线车型

SAKURA VILLA
Global Top Decoration Extravaganza

樱花墅

ISBN: 978-988-12617-6-2 Size: 285X285 mm Pages: 423

The book includes 32 interior design cases. Each is presented with photos as well as floor plans and design concepts, focusing on the application of various accessories in the interior design. Its cases have used the latest, delicate, harmonized and atmospheric elements, all of which reveal the quality lifestyles. The neoteric and fashionable contents of the unique book keep up with the trends; and with its abundant accessories, it offers readers stylish lifestyles while brings new inspirations for professional designers and design enthusiasts.

本书收录了32个室内设计案例。全书无章节划分，各个项目由实景图辅以平面图以及设计理念，着重展现了各种配饰在室内设计中的应用。本书案例中使用的配饰元素丰富新颖、精致，色调统一大气，无不彰显了高品质的生活方式。内容时尚新颖，紧跟潮流，配饰丰富，独具一格，为读者展示极富格调的生活方式。也将为设计工作者及设计爱好者带来不一样的灵感启示。

联系人：卢积灿 联系人：马孝山
订阅电话：**135 1085 8202** 订阅电话：**139 2381 7653**

artpower@artpower.com.cn

DESIGN
EXPLORES
THE SENSES,
EXPLORES
THE SPACES.

KPMG'S
DANISH HQ

KPMG 丹麦办公室

Design Agency | 3XN

A balance between a thoughtful blending into the surrounding scale and an elegant, yet significant expression has been the driving thought behind Danish practice 3XN's design of the new Copenhagen Headquarters for accountancy and consultancy firm, KPMG. The company is certain that their new building gives them the right settings and environment to achieve future growth.

The KPMG Headquarters located in Copenhagen's upscale district of Frederiksberg, has been designed with the goal of fitting into the local setting as a good neighbor. The architects therefore, approached the challenge by designing a building that in spite of its significant size, elegantly adjusts to the surrounding scale. The cloverleaf shape and the light natural stone façade blend in smoothly with the adjacent residential and office buildings.

Inside, the cloverleaf shaped design results in three light atriums, each one being the focal point for the surrounding open office spaces. The atriums are stylistically similar, but not identical, making orientation logical and easy. Footbridges across the atriums are not only aesthetically playful, but also are the quickest route through the building. The logistics have been thoroughly planned to achieve KPMG's goals for greater collaboration between the different departments.

KPMG have been very aware of the added value that great architecture creates for a business. Therefore the building has been designed in close collaboration with 3XN Architects. The result is a building that frames the corporate culture of KPMG and gives a better foundation for accomplishing the business goals, explains Director, and Project Manager from KPMG, Søren Sønderholm.

A balance between a thoughtful blending into the surrounding scale and an elegant, yet significant expression.

完美融于周围环境的建筑规模与优雅又有意义的表达形式之间的平衡。

Confidentiality is important for the clients of KPMG and that requires architecture which builds in discretion and security, while still retaining an atmosphere which is inviting. At the ground level this challenge has been met by a metallic green installation dispersing itself like an inner landscape. These installations are in fact cleverly concealed meeting rooms offering the utmost in discretion, while the exterior landscape is used for the reception and café. Specially commissioned artwork has been etched on the glass walls as a translucent shield to the meeting rooms, and other custom pieces of artwork are evident throughout the rest of the building. Indeed, even in basement parking, colorful photo art gives guests and employees a cheerful welcome as a pre-cursor to the special ambience and architecture of the building.

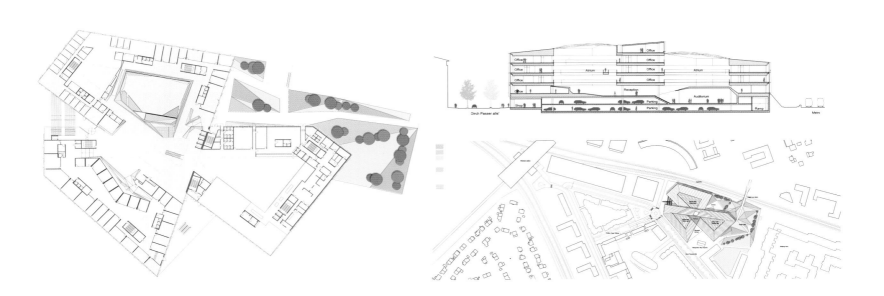

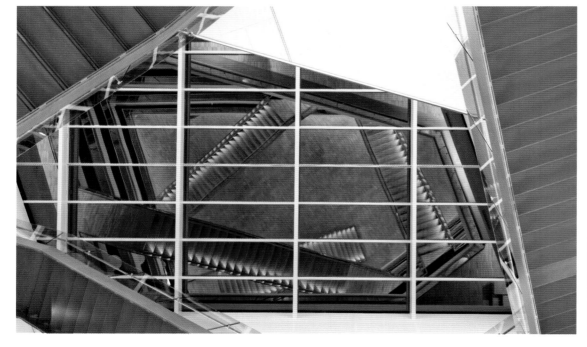

本案为会计与咨询公司——KPMG在哥本哈根的新总部。丹麦设计公司3XN对于该项目的设计理念是：达到完美融于周围环境的建筑规模与优雅又有意义的表达形式之间的平衡。该公司确信，新的办公楼合适的配置与环境能促进企业未来的发展。

KPMG总部位于哥本哈根弗莱德里克堡的高端地区，其设计要融入到周围环境中。设计师在考虑其体量的同时，完美地将其融入了周边建筑之中。三叶草的形状以及浅色天然石块的立面自然地融到周围办公及住宅建筑中。

进入室内，三叶草形状的设计形成了三个明亮的中庭，开放式办公空间以每个中庭为核心展开。三个中庭相似但又不完全一样，易于定位，不至于逻辑混乱。横跨中庭的步行桥不仅形状优美，同时也是穿越建筑物最快速的方式。为了让KPMG公司各部门间能更好地协作，后勤部门在空间中的配置也是经过深思熟虑的。

KPMG深知，一栋优秀的建筑能为商业增加额外的价值。因此，在设计中，通过与3XN的建筑师们紧密合作，造就了这一既能体现KPMG企业文化又能更好地实现KPMG商业目标的建筑体。KPMG的项目经理Søren Sønderholm如是说。

KPMG的客户非常看重机密性，因此建筑本身需给人以谨慎且隐私能得到保障的印象，同时兼有欢迎客户到访的气氛。首层金属质感的绿色装置将自身隔离成一个内部景观。这些装置巧妙地将会议室隐藏其中，最大限度地给人以谨慎私密的感觉。外部则用于接待及休闲。特别定制的艺术作品被蚀刻在半透明的会议室玻璃隔断墙上。整栋建筑中，各种定制艺术作品随处可见。即使是在地下停车场，各种生动的图片似乎也在为这栋建筑特别的氛围预热，向来访的客户及员工们致以热烈的欢迎。

FROYO YOGURTERIA
IN VOLOS, BRANDING DESIGN APPLICATION

沃洛斯市 Froyo Yogurteria 店面品牌设计

Design Agency | Ahylo Studio   Architects | Pavlos Xanthopoulos, Ioulietta Zindrou   Architectural Assistant | Daphne Lada   Branding/Graphic Design | Asprimera Design Studio
Location | Volos, Greece   Area | 70m²   Photographer | Courtesy of Ahylo Studio

Froyo Yogurteria in Volos, is an application of the architectural branding, held for the Greek frozen yogurt franchise chain, Froyo Yogurteria. The main design principle was to highlight the product qualities by distinguishing its identity characteristics. A unique architectural environment aiming to frame the product in a way that would choreograph a free form living environment, and not imitate its icon. The freshness, the color and the texture of yogurt, are reflected on the treatment of the forms, and the choice of the materials and colors.

The yogurt toppings counter and the ceiling of the store are the two elements selected to embed the design logic. Especially, the yogurt toppings counter is the main design element that is repeated in all the stores of the chain. Its clean and imposing form successfully highlights the colorful yogurt toppings, makes the counter recognizable among other frozen yogurt bars, and works as a key element for the brand.

Apart from the counter for the yogurt toppings, which has an adjustable – unique form in all the Froyo Yogurteria stores, the designers chose to emphasize the spatial identity of the store with a double curvature ceiling, painted with the fresh green color of the brand. The resulted inverted, customised, froyo yogurteria green landscape, envelopes the fresh products with reference on yogurt's natural origin.

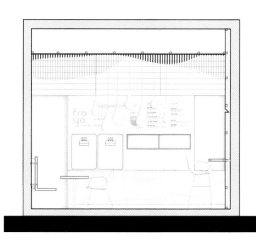

Highlight the product qualities by distinguishing its identity characteristics.
通过凸显品牌特征来强调产品的品质。

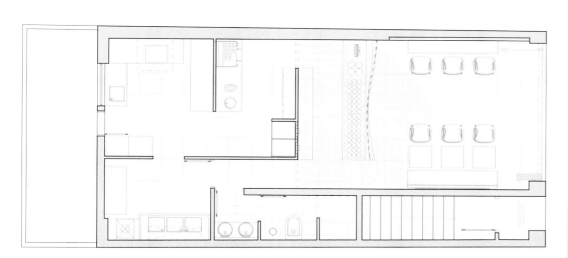

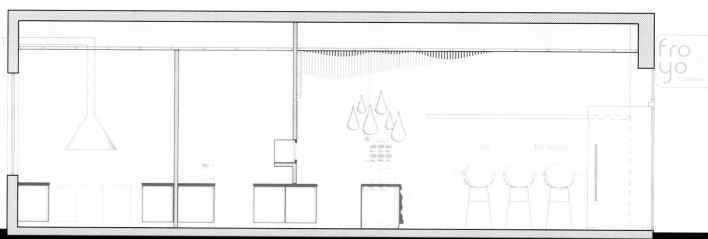

本案是为位于希腊沃洛斯市的一家冻酸奶连锁店面 Froyo Yogurteria 进行的品牌设计。主要的设计理念是通过凸显其品牌特征来强调其产品的品质。独特的建筑形式旨在一定程度上体现其产品并设计出一个自由形式的环境，而不仅仅是简单地运用其商标。酸奶的新鲜特质、色彩及质感，在这一建筑形式的处理、色彩及材质的选择中都得到了体现。

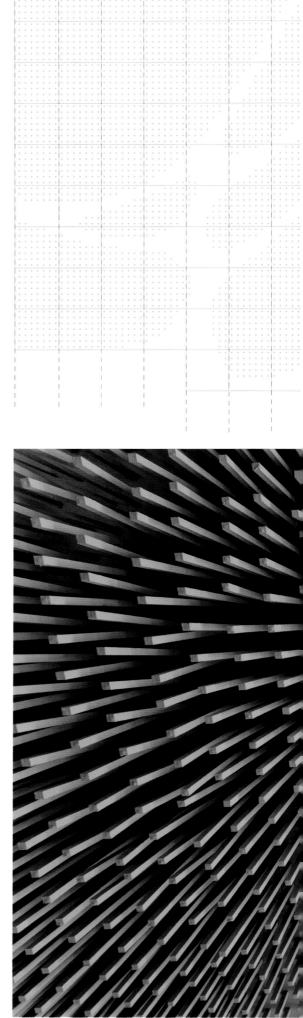

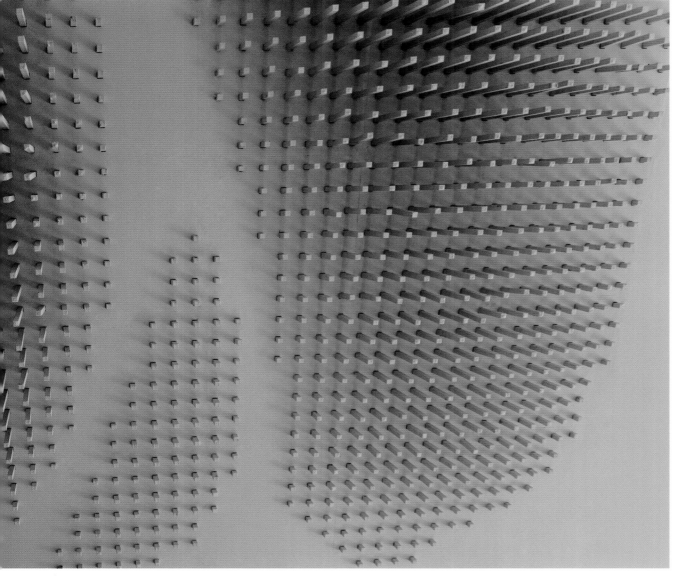

为了体现设计的逻辑，对放置酸奶浇头的柜台以及店内的天花板造型进行了独特的设计。特别值得一提的是，放置酸奶浇头的柜台作为主要元素，被运用在该酸奶品牌的所有连锁店面中。干净且让人印象深刻的柜台造型极具辨识度，且成功地凸显了各种颜色的酸奶浇头，因此被选作该品牌形象的主要元素。

除了放置酸奶浇头的柜台，这间店面做了一些有别于该品牌其他店面的调整。设计师使用涂成嫩绿色的双曲率天花板，加强了该店面的空间品牌形象。特别定制的倒置天花板营造出 Froyo Yogurteria 的绿色景观，将新鲜的产品包覆其中，同时也体现了 Froyo Yogurteria 产品皆使用源于自然的原料。

SPACE | ARTPOWER CREATIVE SPACE

MUMAC
MUSEUM OF COFFEE MACHINE

MuMAC 咖啡机博物馆

Head Designers | Paolo Balzanelli, Valerio Cometti   **Structural Engineering** | Francesco Terreni   **MEP Engineering** | Antonio Bozino   **Client** | Gruppo Cimbali   **Collaborators** | Karolina Kolodziej, Massimo Lapenna, Roberto Lamanna   **Location** | Milan, Italy   **Area** | 1,800m²   **Year** | October 2012   **Photographer** | Angelo Margutti & Associati

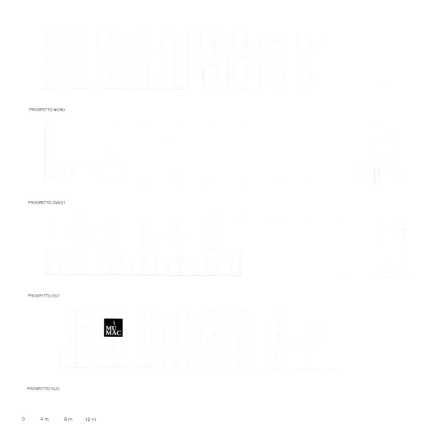

"LaCimbali red" slats of a composite material create a sinuous embrace that has been inspired by the flow of the aroma lifting off a coffee cup and at night a carefully designed illumination creates a strikingly backlit grid of light that evokes the energy living inside MuMAC.

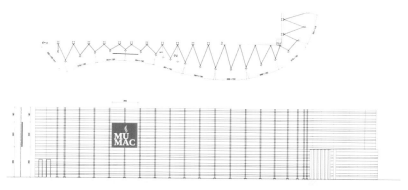

Type Spaces
字体在空间的运用

ISBN: 978-981-07-7383-0   Size: 215X285 mm   Pages: 251

Typography is one of the most intriguing element that pops up all over our daily life. This book explores how we interact with and interpret typography when it is no longer restricted to print or screen. When typography enters three-dimensional space, we can interact with it in ways beyond what the page or screen allows. Bearing that in mind, "Type Spaces" is a book that recognizes this importance that never goes out date and it explores the new meanings that become apparent in text when we can touch it or inhabit it – when image, form, and language converge.

　　字体排列是参与我们日常生活的最有趣的元素之一。本书旨在研究脱离了打印及屏幕后的字体排列与人们的交互影响。当字体进入三维空间，我们可以越过纸张及屏幕与其进行互动。《字体在空间的运用》是一本意识到这一重要性的书，探索了一种新的形式，图像、形状及语言聚集在一起，我们可以直观地触摸到它。

联 系 人：卢积灿	联 系 人：马孝山
订阅电话：**135 1085 8202**	订阅电话：**139 2381 7653**

artpower@artpower.com.cn

SPACE | ARTPOWER CREATIVE SPACE

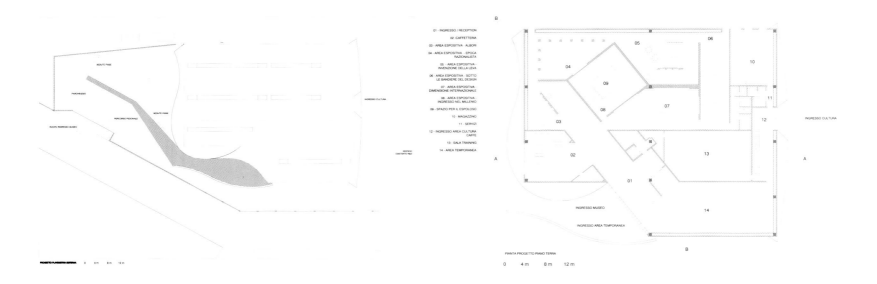

> 蜿蜒的"金佰利红"复合材料板条，灵感来自于咖啡杯散发出的芳香；夜间，精心设计的照明创造出的背光式格栅，唤起 MuMAC 博物馆内部的活力。

MuMAC, Museum of Coffee Machine, was designed by Paolo Balzanelli, owner of Arkispazio, and Valerio Cometti, founder of Valerio Cometti+V12 Design in order to celebrate the 100th anniversary of Cimbali Group, the most important professional coffee machine manufacturer in the world, through its legendary brands LaCimbali and Faema.

MuMAC lies within the establishment of the Cimbali Group a few miles South of Milan.

The core of the new architecture is a building previously used as a warehouse, within which are located both the exposition area and a versatile open space suitable for events and exhibitions to the culture of coffee.

MuMAC narrates a story that spans across 100 years, recounting the history of this extraordinary object within its 1,800 square meters.

Both facades of the museum have been rendered with a delicate nonetheless technological technique: "LaCimbali red" slats of a composite material create a sinuous embrace that has been inspired by the flow of the aroma lifting off a coffee cup and at night a carefully designed illumination creates a strikingly backlit grid of light that evokes the energy living inside MuMAC.

Visitors of MuMAC, access via a new entrance: a coffee-colored wall identifies clearly such opening.

The internal garden is limited by this coffee-colored wall that is marked by nine trees which divide it into 10 equal spaces: ten decades of the century that symbolize the life and the achievements of Cimbali Group.

MuMAC 咖啡机博物馆，是由 Arkispazio 的创始人 Paolo Balzanelli 和 Valerio Cometti+V12 Design 的创始人 Valerio Cometti 为庆祝金佰利集团 100 周年而设计建造的。金佰利集团是世界上最专业的咖啡机生产商，旗下咖啡机品牌有金佰利和飞马。

MuMAC 坐落在金佰利集团范围内，米兰以南几英里处。

该建筑的核心部位以前是一个仓库，内部设立了展示区和开放区，开放区可用来进行一些讲解咖啡文化的活动和展览。

MuMAC 讲述了一段长达 100 多年的故事，在 1800 平方米的空间重述了这一杰出集团的历史。

博物馆的两个立面使用了精湛的工艺：蜿蜒的"金佰利红"复合材料板条，灵感来自于咖啡杯散发出的芳香；夜间，精心设计的照明创造出的背光式格栅，唤起 MuMAC 博物馆内部的活力。

游客们通过一个新的入口进入 MuMAC，一面咖啡色的墙清楚地标志着入口。

内部花园的一侧以这面咖啡色墙为界，九棵树将空间十等分，代表着金佰利集团这个世纪每个十年的历史和成就。

The museum area offers an exposition divided into six historical periods from the beginning of the century to the present day: The early years, The age of Rationalism, Invention of the lever, Under the banners of design, The International dimension and The New Millennium.

Area called The early years is characterized by a suspended ceiling and posters from Art Deco period.

Exhibition of The age of Rationalism include a severe fascist colonnade and strict grid of orthogonal lines identifies the layout of the marble display stands.

In the area dedicated to the culture of the 1950s and 1960s we can find a reconstruction of a bar and an entirely cantilevered structure that supports the machines of the period where thanks to cleverly positioned mirrors, visitors can enjoy both sides of these wonderful machines.

Under the banners of design is characterized by a collection of design masterpieces of the late 1960s and 1970s: great Masters of design have penned coffee machines in these decades, therefore these machines are actual design icons.

The New Millennium is where the display stands are coated in white resin and they smoothly emerge from the floor equally coated in white resin: this area portrays the most modern machines, those are designed for an increasingly fast society. This room has a full height red volume, which is visible from any angle of the museum. Within this volume is visible an installation of the new LaCimbali M100 coffee machine designed by Valerio Cometti+V12 Design: a daring exploded view that allows to grasp the technological content and the level of complexity of such machine, becoming an invitation to reflect on the extraordinary journey that the coffee machine has made during these last hundred years.

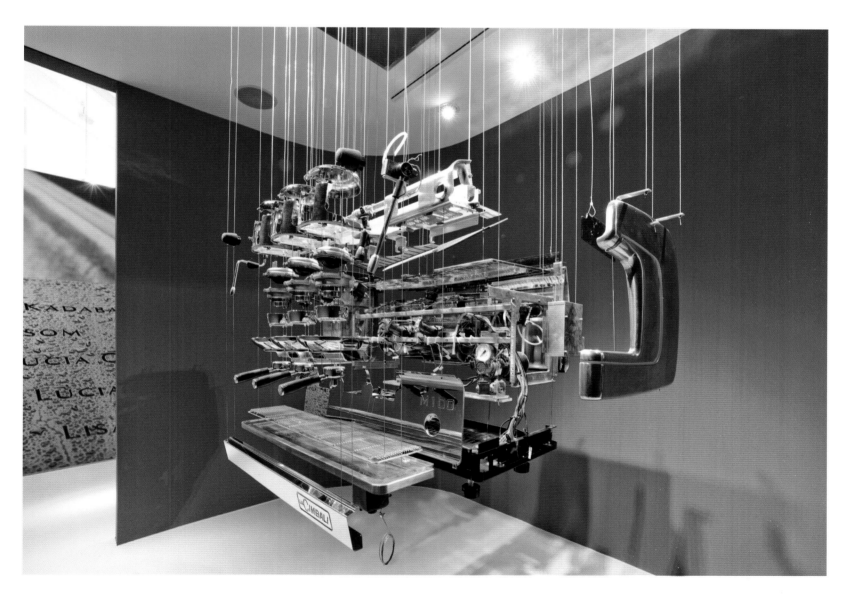

博物馆的展示区分为六部分，分别展示从创立之初到如今六个历史时期的产品。这六个时期分别是：创立之初、理性时期、发明创作期、以设计之名、国际化和新世纪。

创立之初区域的特点是悬浮的天花板和艺术装饰时期的海报。

理性时期展示区包括毫无华饰的法西斯风格的柱廊和标准格栅的线条勾勒出的大理石陈列架。

在20世纪50年代和60年代的展示区域，可以看到一个重建的吧台和一个用于放置这一时期咖啡机的悬臂式结构。镜子的巧妙运用使得游客能欣赏到这些咖啡机的正反两面。

以设计之名区域的特点是20世纪60年代晚期和70年代设计出的一系列咖啡机杰作：这一时期的咖啡机融入了经典的设计，因此这些机器无疑是真正的设计标志。

新世纪区域的展示台都涂上了白色的树脂，从同样覆有白色树脂的地板上平稳地浮现。这一区域展示了为快速发展的社会设计出的最现代的机器。在博物馆内任意角度都能看到这个由楼面至天花板的红色体量。在这一体量内可以看到Valerio Cometti+V12 Design设计的LaCimbali M100咖啡机，大胆地展示了机器的工艺以及这一水准的机器的复杂度，邀请人们见证在这过去的百年间咖啡机发展的历史。

百特普雷斯

Designer | Francisco Sarria

> Use green interior architecture to give expression to green technology.

用绿色的室内建筑来体现绿色工艺。

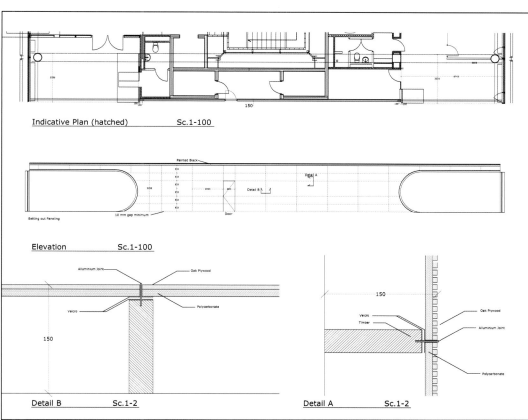

Indicative Plan (hatched)   Sc.1-100

Elevation   Sc.1-100

Detail B   Sc.1-2   Detail A   Sc.1-2

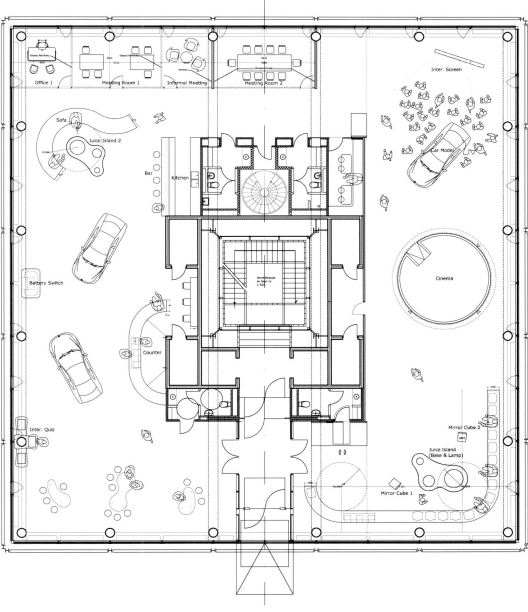

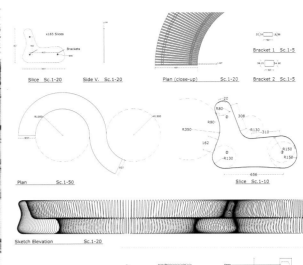

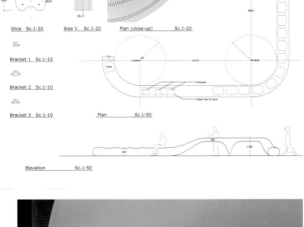

The permanent exhibition of over 1,000 m² conveys the core values of Better Place: namely that innovation and sustainability can go hand in hand and that zero-emission vehicle powered by electricity from renewable sources is the best way forward.

The main concept of the visitor centre is the use of green technology shaped by green interior architecture. The exhibition is a spectacular and engaging experience that aims to entertain and surprise rather than just inform its audience. The space was designed to create an emotive trip that invites visitors to explore, play and learn.

这个1000多平方米的常设展馆表达了百特普雷斯公司的核心价值观：创新和持续性可以并行不悖，使用再生能源电驱的零排放交通工具是最好的解决方式。

这一游客中心的主要设计理念是用绿色的室内建筑来体现绿色工艺。这个展览将提供一次壮观迷人的经历，旨在给游客带来欢乐和惊喜，而不仅仅是传递信息。空间的设计旨在创造一段精彩的旅程，邀请游客来探索、游乐、学习。

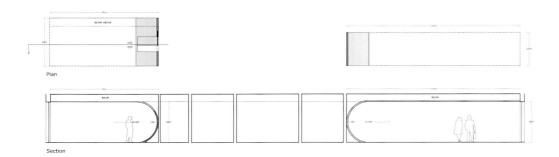

Using camera tracking and directional sound, interactive mirror cubes change from mirror mode to screen mode displaying animated content about the use of electric vehicles.

A cylindrical cinema for up to 20 people is one of the main features of the space. The cinema is operated by a wire system which hoists the cylindrical wall up and down. The cinema and all the audio visual elements of the exhibition are operated remotely.

使用摄像机跟踪和定向声音,将交互式镜面立柱从镜面模式转换到平面模式,播放指导人们使用电驱汽车的动画。

一个能容纳20人的圆柱形影院是这个空间的主要特色。影院由有线系统操作,能控制圆柱体墙面的上下移动。影院以及整个展览中所有视听系统都是远程操控的。

Interactive quizzes to test the guests' knowledge after touring the visitor centre. Touch screens and models illustrating the components of the Better Place infrastructure are all part of the exhibition.

交互式小测试,用于测试游客游览完整个空间后学到的知识。触摸屏和百特普雷斯基础设施的模型图解也是整个展览的一部分。

Apart from conveying the Better Place messages, the installations and interior architecture have a functional purpose, providing furniture and spaces for visitors to meet, play around or just relax.

除了展示百特普雷斯的相关信息，这些装置和室内建筑还有其功能性用途，充当着游客会面、玩乐和放松的设施。

SPACE | ARTPOWER CREATIVE SPACE

WASHINGTON, D.C.

华盛顿特区

Design Agency | HENSE

> **The nature of creating public art is that you are dealing with many different feelings and opinions on art and that can be very subjective.**
>
> 创作一件公共艺术品的本质就是和各种关于艺术的情感以及言论打交道，而这些都是相当主观的。

The project in Washington, D.C. was a fun one. HENSE worked with a small crew to complete it. It was a private commission which was located in SW Washington, D.C. across the street from the Rubell's proposed Contemporary Art Museum. The area in Washington is a part of town that has huge potential to be the next arts district and this project is the first step in bringing some life and color into the area. Taking an existing building like the church and painting the entire thing recontextualizes it and makes it a sculptural object. The designers really wanted to turn the church into a three-dimensional piece of artwork. With projects like this one, the designers really try to use the existing architecture as inspiration for the direction of the painting. HENSE also wanted to use very bright and bold colors to catch a viewer's attention from far away. Most of his works are done in layers and he is never afraid to change the image. The first step was to just get paint and color on every side and surface of the building. The designers then started developing large shapes and gestures that would take days to paint. The entire process took several weeks of layering and working.

HENSE received mostly positive reactions from people there in Washington, D.C. who came to see the progress in person. There were a few people who thought of it as desecrating on the church. Although once it was explained that it was a work in progress and had positive thought behind the gestures, colors and marks, they generally understood.

本案是华盛顿特区一个很有意思的项目。HENSE 和一个小团队一起完成了这个项目。本案位于华盛顿特区西南部，对面是当代艺术博物馆。这一区域拥有巨大的潜力，或将成为下一个艺术特区，本案就是将活力和色彩带入这一区域的第一步。在一座教堂表面绘画，令其焕发新生，仿若雕塑一般。设计师想将这间教堂变成一件三维的艺术品。像这样的项目，设计师们尝试用现有的建筑作为绘画的灵感。HENSE 想要使用明亮大胆的颜色，让人在远处即能看到这栋建筑，并被它吸引。他的大部分作品都经过层层涂绘，他从不怕改变。第一步是在建筑的每一面都涂上色。然后确定大的形状和图案形态，这一步需要几天时间。整个绘制过程需要好几周的时间。

该项目得到了大部分来观看绘画进展的人的积极评价。然而，还是有一部分人认为此举亵渎了教堂。当 HENSE 解释说，这是尚在进展中的工作，这些图案、色彩和标记背后有着积极的思想，人们通常会表示理解。

SPACE | ARTPOWER CREATIVE SPACE

SPACE | ARTPOWER CREATIVE SPACE

PEDESTRIAN BRIDGE

人行天桥

Architect | HHD_FUN   Interior | HHD_FUN   LDI | HHD & XinChao Design   Civil Engineer | H&J International
Façade Consultant | FUDA   Photographer | Zhenfei Wang

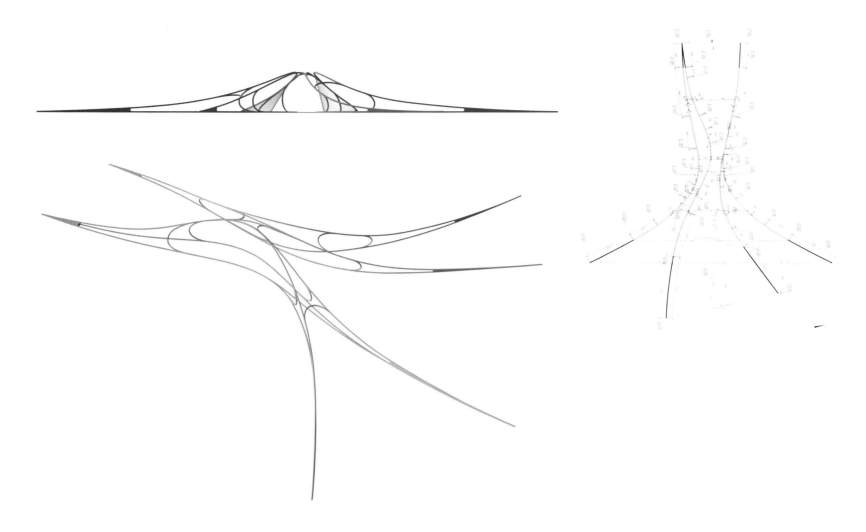

Minimize the construction impact to the natural environment.
最大限度降低建筑对于自然环境的影响。

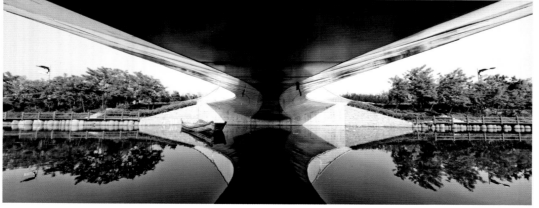

The pedestrian bridge was located on the south part of the 2km long Shanhaitian beach park in Rizhao, east China.

Alongside the beach, the key feature of this park is its 50-year-old black pine forest and the design challenges are to minimize the construction impact to the natural environment. The curved form was strategically designed to allow the 45 meters long bridge fit into the natural environment while at the same time provide a connection between the city and the beach park.

本案位于华东地区日照市 2 公里长的山海天海滩公园南部。

沿着海岸线，该公园的主要特点是树龄达 50 年的黑松森林。本案设计的挑战在于最大限度降低建筑对于自然环境的影响。扭曲的形状旨在让这一 45 米长的天桥融入到自然环境中去，同时在城市及海滩公园间建立联系。

SPACE | ARTPOWER CREATIVE SPACE

SHOFFICE

棚屋办公室

Architect | Platform 5 Architects **Structural Engineer** | Morph Structures **Contractor** | Millimetre **Photographer** | Alan Williams Photography

Smart office space.
奇妙的办公空间。

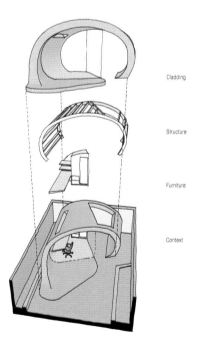

shed + office = Shoffice

"Shoffice" (shed + office) is a garden pavilion containing a small office. The brief required the shoffice to be conceived of as a sculptural object that flowed into the garden space.

A glazed office space nestles into an extruded timber elliptical shell, which curls over itself like a wood shaving, and forms a small terrace in the lawn. The interior is oak lined and fitted out with a storage and a cantilevered desk. Two roof lights – one glazed above the desk with another open to the sky outside the office bring light into the work space.

The project was a close collaboration between Architect, Structural Engineer and Contractor. The lightweight structure, formed with two steel ring beams, timber ribs and a stressed plywood skin, sits on minimal pad foundations. Much of the project was prefabricated to reduce the amount of material.

本案是一个花园小亭子，里面是一个小型的办公室。它仿若雕塑般融入这一花园空间。

光滑的办公空间依偎在木制椭圆形壳中，整体卷曲宛如一个木制的刨花，并在草坪上形成了一个小小的平台。室内是橡木的，并配备了存储空间以及悬臂式桌子。室内有两个采光天窗，一个是在桌子上方的玻璃天窗，另一个则位于办公室外，将光线引入室内。

这个项目是经过建筑师、结构工程师和承建商通力合作完成的。轻质的结构，由两个钢环梁固定成型，木条及胶合板表皮组成最小限度的垫式基础。该项目的大部分都是预制的，以减少材料的使用量。

SPACE | ARTPOWER CREATIVE SPACE

TONY'S FARM

托尼的农场

Design Agency | playze   Client | Tony's Farm   Location | Shanghai, China   Built Area | 1,060m²

A truly sustainable building.
一栋真正的可持续性建筑。

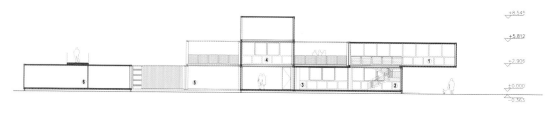

01 Section 01 剖面
1. Tea lounge 休息室
2. Double height entrance 二层通高入口
3. Lobby 大堂
4. Triple height lobby 三层通高大堂
5. Courtyard 庭院
6. Technical room 高科技实验室

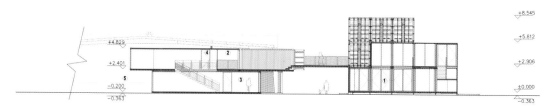

02 Section 02 剖面
1. Double height entrance 二层通高入口
2. Terrace 露台
3. Entrance to office 办公室入口
4. Entrance to exhibition room 展览室入口
5. Warehouse interior 仓库内

Tony's Farm is the biggest organic food farm in Shanghai, which produces OFDC certified (member of IFOAM) vegetables and fruits. But Tony's Farm is meant to be more than just a place for vegetable production. The vision is to integrate the consumer and therefore promote a natural lifestyle.

To link the activities of the working people with the visitors of the farm, playze developed a building complex, which combines the main reception, a lobby, and a vip area, with the new offices and an existing warehouse, where the fruits and vegetables are being packed. The building provides transparency within the manufacturing process. Thus it supports the vision of integrating the visitor and helps to reinforce the consumer confidence in the products of the farm.

The building has been designed as a continuous spatial sequence in order to physically and visually connects various interior and exterior programs. The systemic nature of the containers is countered with the adaptation to the specific situations, like entrance, courtyard, office wing, terraces, etc. The different orientations towards the landscape of the farm, the functional requirements and the spatial sequence are defining each situation of the layout in a specific way, although the spatial framework is the container with its standardized dimensions.

In order to cope with the high aspirations of the client regarding the protection of the environment, several strategies have been used to reduce the energy consumption of the building. The entire structure is well insulated, even though the containers appear in it's raw form. The original container doors have been perforated and serve as external shading blinds at the sun exposed facades to minimize solar heat gain. A geothermal heat pump delivers energy for the air conditioning and floor heating systems. Controlled ventilation helps to optimize air exchange rates and therefore to minimize the energy loss through uncontrolled aeration. The use of LED lighting reduces the general electricity consumption.

Another ambition of the project is to reduce the energy hidden in construction materials, the so called grey energy. Therefore recycled, ecologically sustainable, fast growing or at least recyclable materials have been used. The re-use of freight containers seemed adequate, first for its inherent structural autarky and second for being a common metaphor for "recycled space". Further, the minimal weight of the container structure allowed to re-use the existing foundation plate. The use of local bamboo products for indoor and outdoor flooring, as well as all the built-in furniture additionally supports the ambition of constructing a truly sustainable building.

　　托尼的农场是上海最大的有机食品农场，生产有机食品发展中心（OFDC）认证的蔬菜水果。但托尼的农场不仅仅是一个生产果蔬的场所，而是要和消费者联在一起，提倡一种自然的生活方式。

　　为了将游客和工人们的活动联系起来，playze 设计了一栋综合建筑，包括主要接待区、大厅、VIP 区域、新的办公楼以及现有的供果蔬打包的仓库。这栋建筑可以将生产过程透明化。游客可以直观地看到生产过程，这样可以增强消费者对于农场产品的信任度。

　　这栋建筑被设计成一个连续的空间序列，在功能和视觉上将各个室内外区域联系起来。设计中要考虑到集装箱的自然属性及其被使用的各种场景，比如说入口、庭院、办公室侧翼、阶梯等。尽管空间的框架是标准尺寸的集装箱，但也要详尽考虑各种用途下的布置，均要有利于农场的景观，满足功能的需求，以及空间序列。

　　为了满足客户对于环保方面的高要求，设计师采取了很多措施来降低建筑物的能源消耗。整个结构具有很好的隔热性，即使集装箱看起来是原始的样子。原本的集装箱门被打孔，用于受阳光照射的立面外部，充当遮蔽用的百叶窗，以此来减少因日照产生的热量。地热泵为空调和地暖系统提供能量。可控的通风设备有助于优化空气交换率，因而减少不受控制的换气带来的能量损耗。LED 照明系统的使用降低了整体的电能消耗。

　　该项目的另一项节能措施是减少建筑材料的能耗，即所谓的灰色能量。建筑中使用了回收利用的、生态可持续的、能快速生长的或至少可再次回收利用的材料。货物集装箱的再利用相当合适，一是因为其固有的结构，二是其可以称之为"空间的回收利用"。另外，集装箱结构极小的重量便于在现有的支承板上使用。用当地竹子制成的室内外地板，以及所有的嵌入式家具进一步保证了建造出一个真正的可持续性建筑。

SHREWSB
INTERNATIONAL SCHOOL

曼谷什鲁斯伯里国际学校

Architect | Shma   Location | Bangkok, Thailand   Area | 900m²

This "Play Field" was proposed for a new kindergarten playground for Shrewsbury International School. The existing playground was "defined" with traditional all-in-one play equipment, which ways of play is limited and does not much encourage child brain development. Aiming to break through this typicality, the architect introduces a more appropriate play space for young.

The new playground should enhance their exploration senses and social interaction which would assists child development. The design offers a generous and flexible "Play Field" that can accommodate a range of outdoor class activities. Various "Play Pods" are placed randomly on the field creating a more intimate space with different character defined by the shape of 1.2m high wooden partitions. These comprises of "Swirl Pod" that mimics the experience of moving though snail shell, "Island Pod" which encircled with water plant, "Herb Pod" to develop smell sense with Thai Herbs, "Classroom Pod" with roof for gathering

URY

Explore the Senses, Explore the Spaces.
探索感觉，探索空间。

这一"游乐场地"是为位于泰国曼谷的什鲁斯伯里国际学校所设计的运动场。现有的运动场拥有传统的一体化游乐设施，这些游乐方式是有限的，对于孩子的大脑开发没有太大的促进作用。为了打破这一传统，建筑师设计出了这个更适宜的游乐空间。

新的运动场应该要加强孩子们的探索意识以及社会互动能力，这些有助于他们的成长。该设计提供了广阔灵活的"游乐场地"可以适应各种户外教学活动。各种"游戏豆荚"被随机安置在场地上，1.2米高的木制格栅创造出拥有独特特征的更亲密的空间。这些"涡形豆荚"模拟了蜗牛壳，"岛状豆荚"被水生植物围绕着，"药草豆荚"则散发着泰国药草的味道，拥有屋顶的"教室豆荚"可以容纳整个班级的学生，"沙豆荚"内全天都能玩沙。这些"豆荚"都足够高，孩子们可以在其中创造他们自己的世界，老师也可以

of the whole class, and "Sand Pod" for their all-time favorite sand play. The partitions are high enough for children to create their own world yet allowing teachers to see inside for safety. Circular hole at intervals reinforce another layer of space interaction. The third element of the playground is the "Bicycle Track" which meanders between these pods and sometimes cut through the pods to provide fun rides for their bike class.

Green color tint of the rubber floor (EPDM) is chosen to simulate the feeling of playing on a

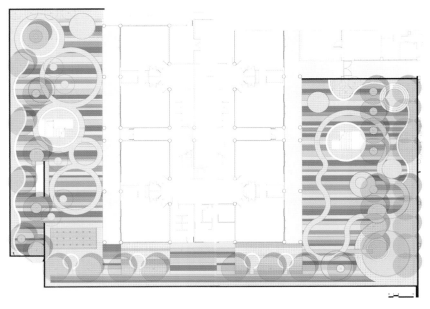

lawn. This also improves the overall environment which is more pleasant to children's sensitive eyes than typical strong primary color flooring. All existing trees are kept for their shade and shrubs are selected for their variety of texture quality highlighted with "Iris" which move softly along with the wind and "Elephant Ear" which feature an upright giant leaf that is easily surpass the height of the pupil. It's an ultimate play garden experience.

By exploring the playground, children are facing with constantly changing spaces from inside to outside, from not being seen to being seen, and from private to public with endless possibility. The Playground will perhaps prepare them the exploring sense and develop their social interaction skill which is vital for their future.

看到里面的情况以确保孩子们的安全。间隔的圆孔在另一层面上加强了空间的交互作用。这个运动场的第三种元素是"自行车"赛道。他们蜿蜒在这些"豆荚"之间,时而穿透这些"豆荚",为孩子们的自行车课程提供了很多骑车的乐趣。

绿色调的橡胶地板模仿出在草坪上玩乐的感觉。与典型的原色地板相比,这一色调的地板可以为孩子们敏感的眼睛提供愉悦的感觉,也同样改善了整体的环境。原有的树木都得以保留,用于遮阳。灌木则根据其质感的多样性来进行选择,"爱丽丝"随风轻柔摆动,"大象的耳朵"拥有直立的大叶子,可以轻易高过孩子们的身高。这是一个终极游乐花园体验。

在探索运动场的过程中,孩子们将面临着从里到外的不断转变,时而能被看到,时而不能,从私密空间到公共空间,充满了无限的可能性。该运动场也许将激发孩子们的探索意识,发展他们的社会互动技能,这些对于他们的将来是至关重要的。

VILLA SSK

SSK 别墅

Design Agency | Takeshi Hirobe Architects **Architect** | Takeshi Hirobe **Structural Engineer** | Akira Ouchi /S FORM **Location** | Chiba, Japan **Structure** | wooden structure
Site Area | 854.28m² **Total Floor Area** | 105.05m² **Photographer** | Koichi Torimura

The location for this project was adjacent to the calm and tranquil waters of Tokyo Bay. When first visited the site, the designers were immediately overwhelmed by the presence of the ocean. Soon, however, they became attuned to the subtle cues coming from the surroundings, such as the movements of the sun, the flow of the wind, the scent of the tide, the stirrings of the plants, and the presence of the rocky mountain. They were particularly impressed by the sea and the soaring verticality of the sky overhead, as well as the rocky mountain on the opposite side and the vegetation growing on them. For this reason, they decided to design a building that would serve to connect these mountains to the imposing expanse of the ocean.

The client requested a comfortable, generously proportioned living, dining and kitchen area, a bathroom overlooking the ocean, a guest room, and a spare room that could be used to display his beloved car. The portion of the house encircled by the central area, guest room and spare room were turned into a tiled central courtyard where dogs can play. The courtyard also serves as an outdoor living room where large numbers of guests can gather and mingle. Water can also

Villa SSK

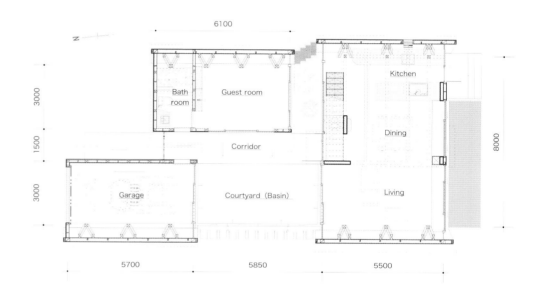

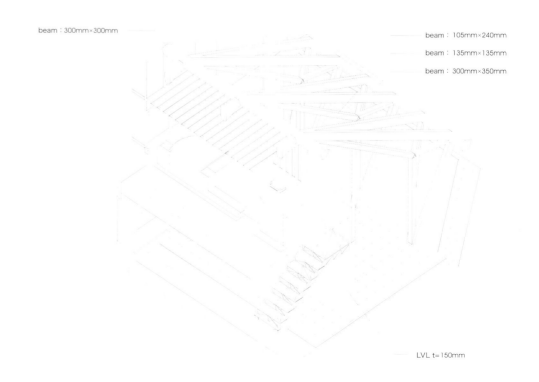

Architecture ought to be rooted in the place it occupies.

建筑要扎根于其所处的土地上。

flow into the courtyard to create a basin that reflects the subtle nuances of the surrounding light and wind. Timber panel wall-building and timber space trusses were used to create a structure that connects the ocean-facing side and mountain-facing side of the house. Although the trusses themselves occupy a wide space, gaps have been left in between the different materials. By angling the walls obliquely in accordance with the necessary spacing and throat gaps required for the truss structure, the designers were able to eliminate certain oppressiveness from the interior.

Architecture ought to be rooted in the place it occupies. The architectural form of this building somehow emerged during the long process of analyzing and studying the location. Although the design process was supposed to have entailed a frantic accumulation of decision-making and choosing between possible options, the finished building gives one the strange, lingering impression of having been constructed according to some law or other.

该项目毗邻风平浪静的东京湾水域。第一次来到项目地点时，设计师立刻就被大海的壮观所倾倒。然而很快，大家就熟悉了周围环境所带来的微妙变化，例如，太阳的移动、风的走向、潮汐的气味、植被的生长，以及落基山脉的存在。给人留下尤其深刻的印象的是壮观的大海、高空直射的阳光、对面的落基山脉及生长在周围的植被。基于此原因，设计师们决定修建一座将壮阔的大海和巍峨的群山联系起来的建筑。

客户要求房子具备如下功能：一间舒适开阔的起居室、餐厅和厨房区域、一间可以远眺大海的卧室、一间客房、一间能用于展示存放户主爱车的空房。房子的一部分被中央区域围绕着，客房和空房则被设计成一个平铺的中央庭院，小狗可以在这里嬉戏。

庭院作为一个户外的起居室，可以招待很多客人同聚。水流至庭院形成一个水池，反射出周围灯光及风的微妙变化。木材面板造壁和木材空间桁架被用来创建一个结构连接该建筑邻近海洋的一侧及靠山的一侧。尽管桁架本身占据了很大空间，然而在不同材质间也留下了众多缝隙。将墙体倾斜一定的角度并预留了桁架所需的空间及缝隙，这样可以消除一定的室内压抑感。

建筑要扎根于其所处的土地上。这栋建筑的建筑形式是在对其位置进行分析和研究的漫长过程中形成的。尽管设计过程需要经过疯狂的决策及各种可行方案的选择，根据一些规律进行建造完工后，该项目呈现给人们的是一种奇特、挥之不去的印象。

SPACE | ARTPOWER CREATIVE SPACE

DIESEL HOME
COLLECTION INSTALLATION

数字——DIESEL 家居展示

Design Agency | Tsutsumi & Associates   Location | Tokyo, Japan   Area | 45m²   Photographer | Kanta Ushio

What on earth does it mean to "show things"? This simple question is the one the designers need to ask themselves in order to realize a display of merchandise and spatial installation at the same time. Generally a room is surrounded by hard substances, to which we fix things in one of the following three ways: to place them on it, hitch them to it, or hang them from it. But a flexible wall with plasticity would give rise to a fourth way, i.e. burying them in it.

Like a cave dug in a cliff, or fossils hundreds of millions years old trapped in lava, when something is buried in another entity, it constitutes an inseparable whole with the background. If the designers apply this to the shop display, they should be able to develop a new display method where goods and space are fused with each other. To create a plastic wall, the designers have devised drawers made small to the limit and piled them in huge numbers. Considering cost, time allowed for the project, and weight, they chose to use paper square pipes 40×40mm for the main body and 35X35mm for the drawers. 10,000 set of drawers with this tiny slide mechanism will be prepared, which will cover the whole wall.

They can be used in any way possible. They could be like glyphs when three-dimensionality is emphasized, like an inlay when flatness is focused, and even "corroded" with light. This is a DIGITAL wall made of tiny squares which people can just DIG IT their own way. Possibilities for display are infinite here, as is our imagination.

展示意味着什么呢？这是设计师们在同时实现展示商品及空间装置之前需要回答的问题。空间中放置物品，通常我们可以用三种方式将物品固定住，即水平放置、挂在墙上或是吊在天花板上。然而，如果有可塑的灵活墙面，则可以有第四种放置物品的方式：把物品埋在其中。

如同悬崖上的洞穴，或是在火山熔岩中存在了亿万年的化石，当一件物品以另一种存在方式被掩埋起来，则物品本身和其所处的背景，就变成了密不可分的整体。如果将这种方式应用到商店陈设中，将创造出一种新的陈设方式，商品和空间完美融合在一起。为了做出可塑的墙面，设计师们设计出非常小的格子，并将其大量堆积在一起。考虑到成本、项目耗时以及重量，设计师们使用了40mm×40mm的正方体纸管作为主体，35mm×35mm的纸管作为格子。大约使用了10 000组这种小小的可以滑动的格子，覆盖在整面墙上。

这些格子可以有各种各样的使用方式。当强调三维性时，它们就像雕刻字形；当强调平整度时，它们又像镶嵌物，甚至像被光线所"蚀刻"。这是一面由小格子组成的数字墙，人们可以用自己的方式来探究。通过发挥想象力，可以让商品的陈设在这里充满无限的可能性。

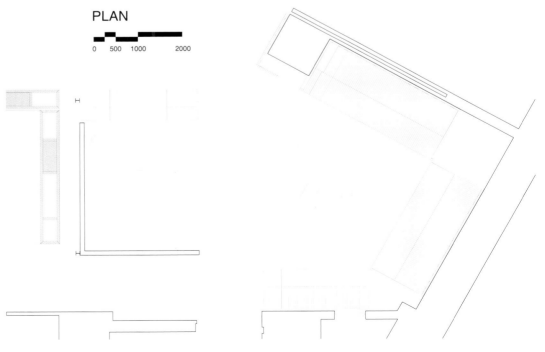

PLAN

Bury them! Create a new display method where goods and space are fused with each other.

埋起来!创造出一种新的陈设方式,将商品和空间完美融合在一起。

THE ARCHITECTURAL DESIGN PROPOSAL SIGNIFIES A DESIRE TO CHALLENGE THE USE OF TECHNOLOGY IN ARCHITECTURE, NOT FOR MERELY CONVENIENCE BUT AN ALL-ROUND ADDITION TO FORM AND FUNCTION.

CONCEPT | ARTPOWER CREATIVE SPACE

AURORA BOREALIS ARCTIC OBSERVATORY — THE WINGS OF THE DAWN GODDESS

北极光天文台——曙光女神的翅膀

Designers | Jensen Liu, Sally Hsu

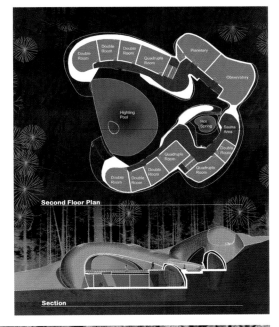

Siivet aurora, the Aurora Borealis Arctic Observatory is an architectural project situated in argument between the historic mystic of the northern lights and the urban landmark of Rovaniemi. The stories of the aurora phenomenon may now have a scientific explanation, and this has not diminish the fascination for this seasonal event. Nature's most fascinating live painting continues to attract countless trekkers to search for this phenomenal experience, for that the designers have designed a hotel to capture and enrich this priceless journey.

Through years of evolution the architectural design and construction has yet to proceed greatly in the use of technology. The architectural discipline continues to be a step behind the technological advancement of the modern age. This proposal signifies a desire to challenge the use of technology in architecture, not for merely convenience but an all-round addition to form and function. How does architecture embrace the primitive desire of its occupant to immerse in a mysterious fantasy yet also project forward into the future way of living? This proposal explores the awakening moment of the Eos (the goddess of dawn). The awakening enhances a natural phenomenon that continues to amaze audiences.

The project believes sustainability should not be implemented merely due to society's expectation on green architecture. Sustainability encourages precise consideration of the site context. An efficient sustainable system helps minimize the disturbance of the site and more important implemented as a feature that adds to the architecture not just functionally but also formally. The aurora is signifying the goddess of dawn splitting open the night sky breaking free of the dark skyline. The architectural design signifies Eos's wings responding to the natural aurora phenomenon, as the magnetic field

极光翼——北极光天文台是一个建筑设计项目，坐落于在历史上具有神秘色彩的极光城市Rovaniemi。现在，可能已经对极光现象有了自然科学的解释，但这似乎并没有减少人们对这种季节奇景的迷恋，这个大自然最迷人的现场画继续吸引着无数千里跋涉寻找这个惊人现象的旅人。酒店的设计就是希望能捕捉和丰富这无价的旅程体验。

随着多年建筑设计和建造的不断演化，其进程开始很大程度依赖科技的应用。然而建筑学科的发展依然落于摩登时代科技进步之后。这一项目想要展示一种挑战的渴望，将科技应用于建筑，这种应用不仅仅要落实在便利性方面，它应该包括全方位的外在造型及功能。建筑应如何与其居住者的原始欲望相拥抱，让居住者能够沉浸在神秘幻想的同时看到一种属于未来的生活方式？这一项目探索曙光女神最令人惊叹的瞬间，而这一瞬间与自然景观相辅相成，带来更大的震撼。

建筑永续发展的观念不仅仅只是一个社会所期望的绿色建筑的概念，而是鼓励设计师要切实考虑场地环境。一个有效的可持续的系统，有助于最大限度地减少对基地特色的干扰，更重要的是不仅仅提出对功能的可执行性，还有形态上的考虑。曙光女神打开翅膀划破黑暗夜空形成极光；建筑设计模拟曙光女神的翅膀以呼应自然极光现象。由磁力引导及风力鼓动，女神之翼在天际的极光中自然摆动，构成一幅自然界中的绝美画作。

建筑外层由反光的不锈钢构成，使用这种反射

CONCEPT | ARTPOWER CREATIVE SPACE

and wind undulates, it follows throughout echoing nature's best painting.

The building has chosen reflective stainless steel to be the facade finish. This reflective surface aims to capture the aurora and blends within the context. The architecture sits within the terrain; it minimizes the conflict with the site. This nestling aims to utilize the earth thermal energy to achieve insulation.

The roof top has installed LED light tubes, as the aurora surface, the light tubes will respond to the magnetic forces, creating a spectacle of dance within nature. The architecture nestle within the natural landscape. The building center has a seamless water surface that is able to seamlessly connect with the horizon ahead, creating the sensation of reaching into the sky.

The architecture has installed a lighting pool with the function of a skylight in the entrance foyer of the building. In the cold harsh winter, the visitors and occupants within the interior can still enjoy a panorama scenery and rich daylight. This installment act as insulation, the interior temperature will be able to be maintained for comfort. This ensures the visual experiences do not lessen the comfort within the building.

To ensure the building is sufficient and sustainable, geothermal energy is captured to transform into electricity for the generation of a magnetic field. This power is used to generate the magnetic field required to active the façade of the building. The force field allows the LED wings to undulate respond to the Aurora and the force all around. The water collected from snow and rain will be recycled to be used within the building in all leisure hot springs and water features.

The building site is chosen in the middle axis of the river, to avoid light contamination from the city and provide best viewing opportunities for the visitors.

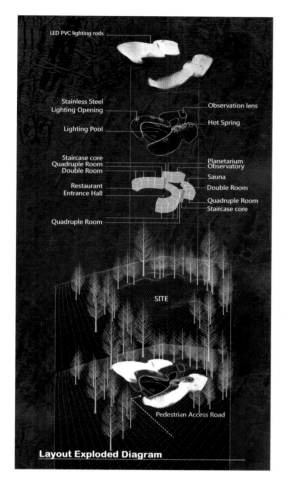

Layout Exploded Diagram

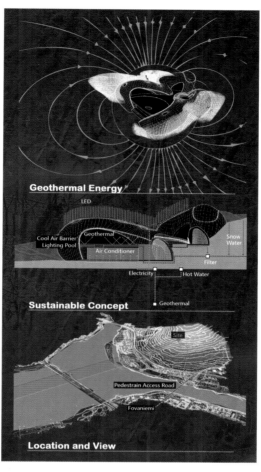

Geothermal Energy

Sustainable Concept

Location and View

表面的目的在于捕捉极光并与周围景致相融合。整个建筑融于地势之中,最大程度地缓和了与环境的冲突。这种巢式设计旨在利用地热能来达到建筑的保温效果。

建筑顶部安装有 LED 灯管,作为极光表面。灯管受磁力影响而摆动,创造出一种随自然而舞的奇景。建筑像鸟巢一样置于自然景观中,建筑中心有一个无边水面,让体验者能在翅膀所包覆的中心随水面的反射体验无边无际的感受。

建筑内建有一个灯光池,充当入口大厅的天窗。在寒冷的冬季,游客和住户在室内仍然可以享受全景风光和丰富的日光。这一装置还能作为隔温材料,使室内保持宜人温度。它在保证视觉体验的同时还达到了舒适度的要求。

为保证建筑的能源供应及可持续性,地热能被转化为电能以产生磁场。磁场是保证建筑外观设计的必需能量,使得 LED 构成的翅膀能够随着极光而舞动。由雨雪收集而来的水资源被回收再利用于室内的休闲温泉用水及水景中。

基地选在拥有最佳视野的河道轴线中央的山坡上,以避免人造光害并创造能延伸极光视野的景象。

CONCEPT | ARTPOWER CREATIVE SPACE

PHOENIX OBSERVATION TOWER

凤凰城观景塔

Design Agency | BIG   **Project Leader** | Iannis Kandyliaris   **Team** | Thomas Fagan, Aaron Hales, Ola Hariri, Dennis Harvey, Beat Schenk   **Client** | Novawest   **Partner in Charge** | Bjarke Ingels and Thomas Christoffersen   **Collaborators** | MKA (structure), Atelier10 (sustainability), Gensler (local architect), TenEyck (landscape)   **Location** | Phoenix, Arizona, USA   **Area** | 6,503m²

BIG is commissioned by Novawest to design a 128m tall mixed-use observation tower to serve as a symbol for the city of Phoenix, Arizona.

Located in downtown Phoenix, the 6,503m² Observation Tower shall add a significant structure to the Phoenix skyline from which to enjoy the city's spectacular views of the surrounding mountain ranges and dramatic sunsets.

The future observation tower is conceived as a tall core of reinforced concrete with an open-air spiral sphere at its top, resembling a metaphorical pin firmly marking a location on a map. The spiraling sphere contains flexible exhibition, retail and recreational spaces which are accessed via three glass elevators that connect the base with the summit and offer panoramic views of the city and the tower's programs as visitors ascend or descend.

Walking downwards from the top through a continuous spiral promenade, the visitors of the observation tower experience all of the building's programs in a constant motion, while enjoying dynamic 360 degree views of the city of Phoenix and the Arizonian landscape.

Once the visitors reach the middle of the sphere, they can choose to either conclude their journey by taking the elevator back to the ground, or continue to the restaurant levels at the lower hemisphere. The motion resembles a journey through the center of a planet, and a travel from the north to the South Pole.

The base of the tower will serve as a public plaza offering shade, water features and a small amount of retail together with a subterranean queuing area. The tower will serve as a working model of sustainable energy practices, incorporating a blend of solar and other technologies.

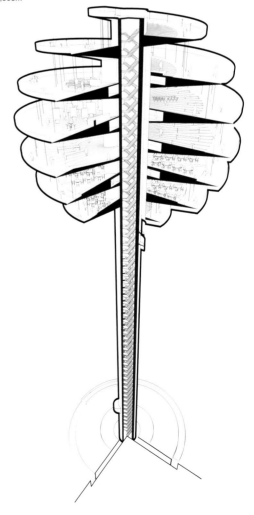

ARTPOWER CREATIVE SPACE | CONCEPT

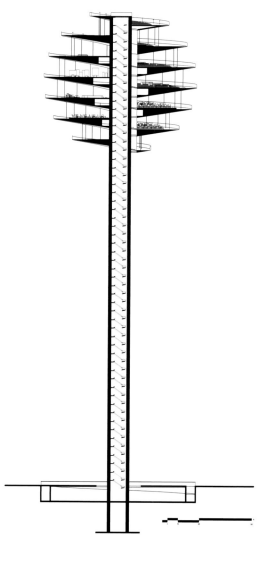

CONCEPT | ARTPOWER CREATIVE SPACE

BIG 受 Novawest 委托设计了亚利桑那州凤凰城的新标志——高达 128 米的多功能观景塔。

这个 6503 平方米的观景台位于凤凰城的市中心,为凤凰城的天际增添了一抹象征性的轮廓,为人们提供了欣赏城市壮观景象、周围山脉和壮美落日的好地方。

这个未来观景塔设计有一个高高的钢筋混凝土核心,顶端有一个露天螺旋形球体,形态上就像一根钉在大都市这张地图上的定位针。螺旋球上包括可灵活应用的展览区、零售店和休闲空间。这些区域能够通过三台玻璃升降电梯到达。电梯接连起了建筑底部和顶部,当它升降时,能够为乘客提供观赏城市全景和塔内各个构筑的机会。

由顶端沿着螺旋长廊漫步而下,观景塔的游客能够在一种匀速运动中体验整个建筑的构筑,同时享受 360 度凤凰城的景色和亚利桑那州景观。

当游客到达球体中部时,他们可以选择搭乘电梯回到地面,以此结束自己的旅程,或者继续前往位于下半球体的餐厅层。这种移动就像从一个星球的中部开始的旅途,或是从北极到南极的远征。

塔底是一个公共广场,有遮阳处、水景和少量带有地下排队区的零售店。这座塔同时能够作为一个运作中的、应用太阳能和其他高新技术的可持续能源使用的范例。

MARINE RESEARCH COMPLEX

海洋研究中心综合体

Designer | Shahira Hammad   **Location** | Alexandria, Egypt

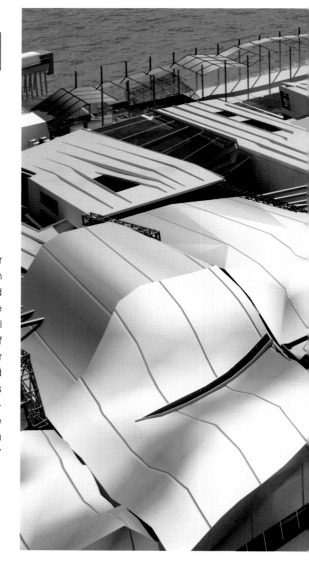

The aim of the project was to create a mixed-use research complex that acts as an integral node of the multidisciplinary field of marine sciences in Alexandria for both public use and research use. Simultaneously this project will increase the city's environmental awareness and help in the conservation of the marine environment.

The complex grounds are located to the east of Abu-Qir's waterfront and cover a surface of 148,000 square meters.

The complex is entirely driven by marine sciences, composed of two key elements: public zone (museum) and research zone (research buildings and fields). Both zones are linked by an outdoor green area, acting as a transitional space between the two main zones. Despite the different use and construction materials of the two zones, both are linked together with a common main structural grid. Extensions projected from the boundaries of each research building were used as guidelines for the museum's grid. An outdoor structure of trussed frames partially shrouded with aluminum panels is built over the transitional space to form a semi-enclosed space, at the same time increasing the cohesion between the two zones. As a result a dovetailing compound of buildings or a "complex" is achieved.

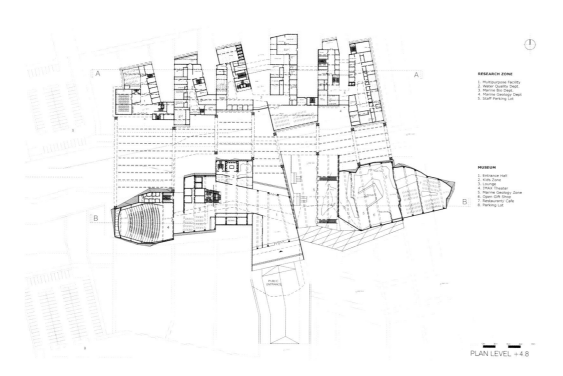

PLAN LEVEL +4.8

这个项目旨在创建一个多功能研究性综合体，用以充当埃及亚历山大港上海洋科学多学科领域的中心，作为公共设施和研究使用。与此同时，这个项目还能够增强城市环境意识，有助于对海洋环境的监测。

这一综合体坐落于阿布吉尔滨水区东部，占地 148 000 平方米。

它完全由海洋科学为主导，包括两个关键区：公共区域（博物馆）和研究区域（研究大楼和试验田）。二者由一片户外绿色区域相连，这块区域成为了两大主区的过渡区域。尽管两个区域在建筑材料和用途上各不相同，但它们通过一条共同的主结构脉络相连。由每栋研究大楼边缘延展出去的领域都被用做指示通往博物馆片区的引导系统。部分覆盖着铝板的户外桁架框架式结构建于过渡空间之上，形成了一个半封闭空间，同时加强了两个区域的联系。如此，一个通过燕尾榫接合方式组合的建筑群或者说"综合体"便建成了。

CONCEPT | ARTPOWER CREATIVE SPACE

South Elevation

North Elevation

ARTPOWER CREATIVE SPACE | CONCEPT

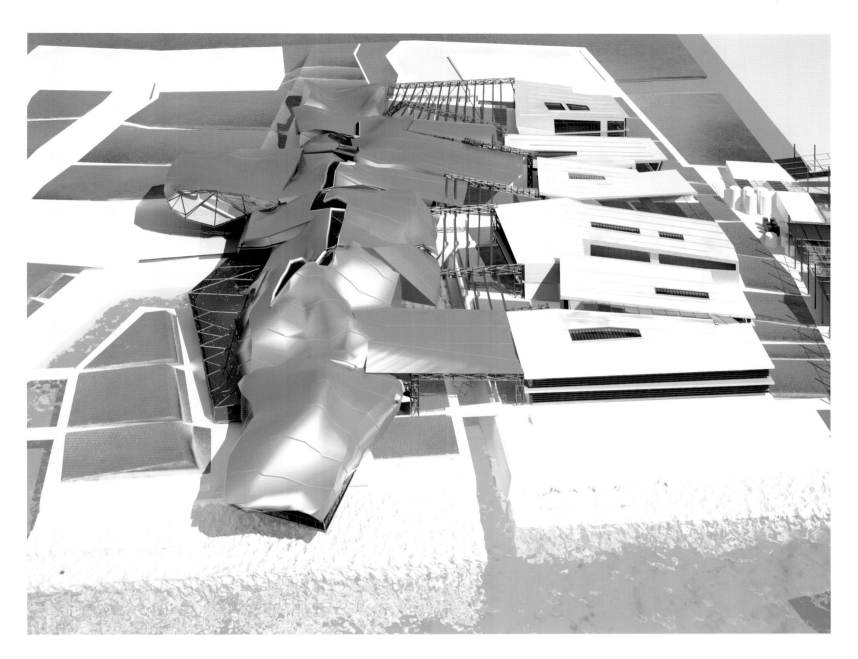

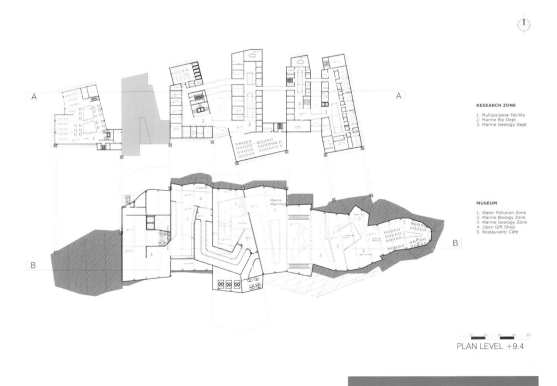

RESEARCH ZONE
1. Multipurpose Facility
2. Marine Bio Dept.
3. Marine Geology Dept.

MUSEUM
1. Water Pollution Zone
2. Marine Biology Zone
3. Marine Geology Zone
4. Open Gift Shop
5. Restaurant/ Cafe

PLAN LEVEL +9.4

WESTBAHNHOF U-BAHN

WESTBAHNHOF 地铁站

Designer | Shahira Hammad   **Location** | Vienna   **Advisors** | Hernan Diaz Alonso, Steven Ma

How to design a subway station?

Should it just be a utilitarian structure, devoid of any attributes besides offering a roof and some walls and a platform or two?

Shahira Hammad decided to make a statement: to "confuse" in order to provoke emotions not normally associated with a subway station.

His early studies were based on the morphologies and behaviors of insects and plants. Through a series of experiments and spontaneous compositions the work evolved into a field — a web of structures, meshes and sculptural elements.

This intricate, complex web tries to create that locus most subways stations are not.

In a way it is the very opposite of an "oasis".

An oasis is a patch of light, serenity and open space.

His subway station is dark, dense, and almost capable of generating claustrophobia.

It is a physical space, but also a psychic space.

It is a labyrinth.

It is an image of his mind, perhaps, living in a world with many contradictions, in a non-linear world, in a world as-a-knot.

如何设计一个地铁车站?

它会仅仅是一个完全实用的结构,除了一个顶盖、一些墙面和一到两个平台外再无其他?

Shahira Hammad 希望能够创造出一种"疑惑感"来唤起人们对地铁车站默认感之外的情感。

他早期的研究基于昆虫和植物的形态学和行为学。通过一系列的实验和自发组合,这项工作发展成了一个领域——一个包括了结构、网眼和雕塑元素的网。

这个错综复杂的网试图将这个车站变为与大多数车站大不相同的所在。

它与"舒适的绿洲"大相径庭。

所谓绿洲是光线充足、宁静的开放空间。

他设计的地铁车站黑暗、浓密,并很有可能让人产生幽闭恐惧症。

它是一个物理性空间,但也是一个精神空间。

它是一个迷宫。

它是他的思想,也许,生存在这个世界,一个非线性的世界,一个如一个绳结一样的世界,总是有着许许多多的矛盾。

CONCEPT | ARTPOWER CREATIVE SPACE

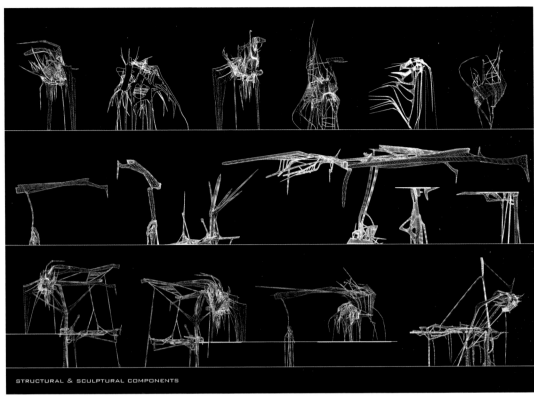

STRUCTURAL & SCULPTURAL COMPONENTS

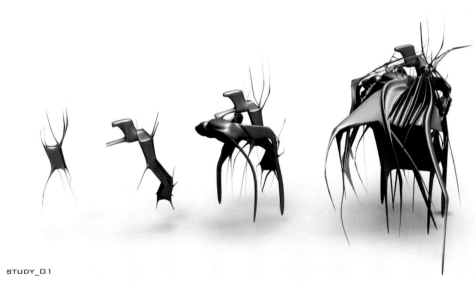

STUDY_01

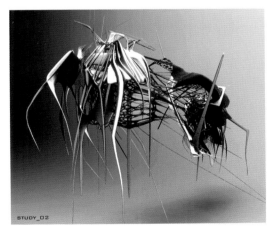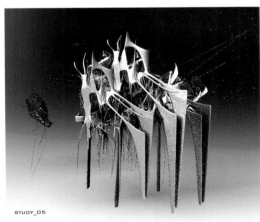

STUDY_02   STUDY_05

COLUMN
专栏

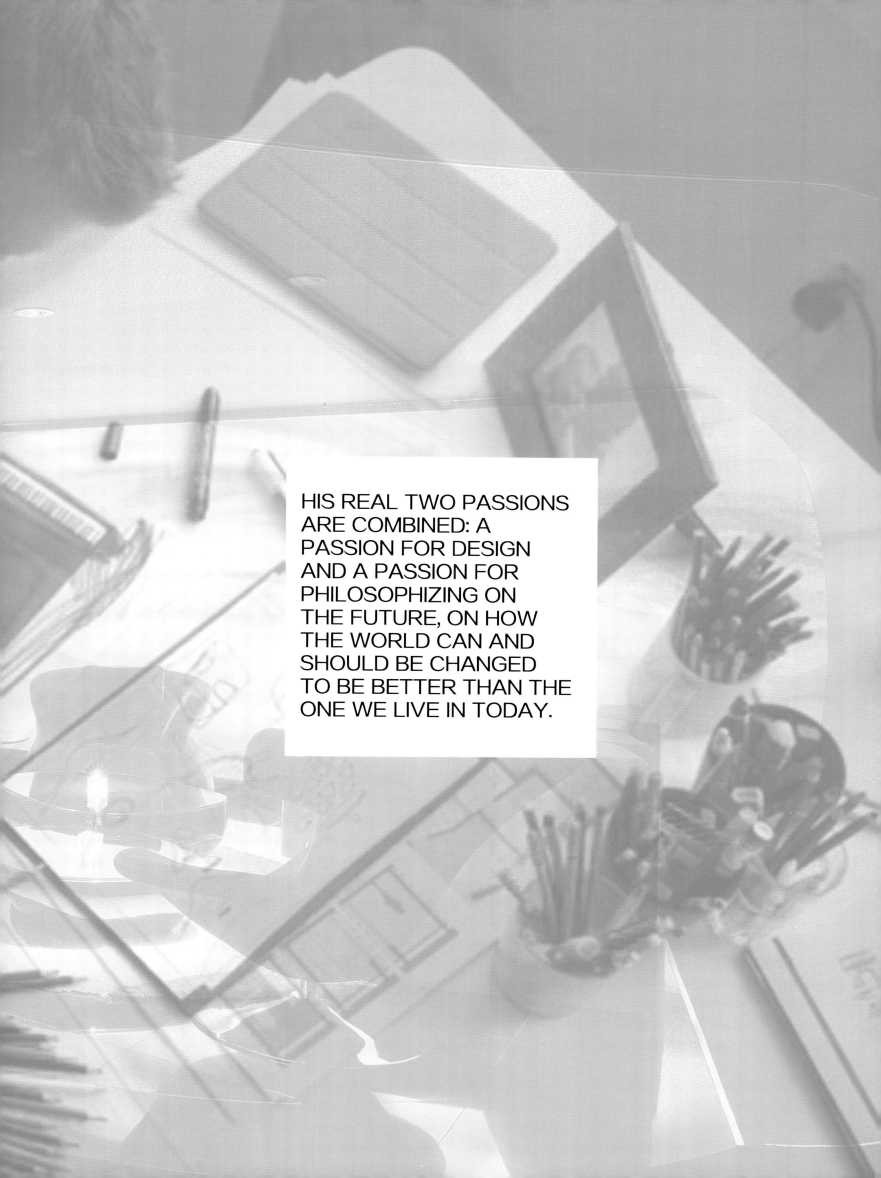

HIS REAL TWO PASSIONS ARE COMBINED: A PASSION FOR DESIGN AND A PASSION FOR PHILOSOPHIZING ON THE FUTURE, ON HOW THE WORLD CAN AND SHOULD BE CHANGED TO BE BETTER THAN THE ONE WE LIVE IN TODAY.

STORIES BEHIND PASSIONS

背后的故事 · 热火设计

Robert Majkut
罗伯特 · 迈酷

Robert Majkut is one of the most important European designers. His hard-earned brand and consistency in his approach to design makes him recognizable as a very popular designer and creator of unique places.

Born in Szczecin, Poland in 1971, he graduated with honours from an art secondary school with furniture design as major. Having first studied architecture at Szczecin University of Technology, he then moved to the Academy of Fine Arts in Poznan. Finally he graduated from Cultural Studies at Poznan University.

Since 1988 he has been involved in art as graphic artist, painter and creator of installations. Natural born rebel, humanist, cultural anthropologist. Today Robert is most of all a designer, effectively changing the reality that surrounds us. For the better.

He represents a conscious level of designing, in which natural, cultural, humanistic and economic aspects are the most important. His style combines formal design with innovation and latest technologies. Because of his education and interests, he always puts his projects in proper cultural and social context.

Robert's work received many accolades – in 2002 he was honoured with the "Rising Star" award by the British Council. He was also nominated for Elle Style Awards 2007 in the "good design" category. In 2011 Polish marketing magazine *Brief* included him in the ranking of 50 most creative people in business.

He has been active in the field of popularising good design. Member of the programme council of New Culture Bec Zmiana Foundation; guest speaker at universities, judge at competitions, expert quoted in Polish and foreign press.

罗伯特·迈酷是欧洲最重要的设计师之一。他以精心打造的品牌和对自己设计方式的坚持成为一位受人肯定的著名设计师和许多独特领域的开创者。

罗伯特1971年出生于波兰什切青市，以优异的成绩毕业于艺术中学家具设计专业。在什切青科技大学，他开始学习建筑，之后转入波兹南美术学院。最后，他在波兹南大学学习文化研究，顺利毕业。

自1988年，他作为一名平面艺术家、画家和装置创作者开始接触艺术。他是天生的反叛者、人文主义者和文化人类学家。如今的罗伯特更多的是一名设计师，正在以自己的能力有效地改变周遭，使之变得更好。

他代表着设计的一个意识层次，在这个层次上自然、文化、人文、经济方面才是最为重要的。他的风格结合了常规设计和新方法、新技术。由于他的教育背景和兴趣所在，他总是将项目放在一个合适的文化和社会背景下。

罗伯特的项目为他赢得了诸多荣誉，包括2002年由英国文化委员会颁发的"新星奖"。他也曾在2007年艾莉风格奖"优秀设计"类别中被提名。2011年波兰营销杂志《Brief》将其评为商界50个最具创意人物之一。

他一直活跃在推广优秀设计的领域，同时还是Bec Zmiana新文化基金计划委员会会员、大学客席讲者、大赛裁判以及众多波兰和国际刊物中所提到的设计专家。

A: You are a designer, humanist, cultural anthropologist, businessman and creator, then how could you keep balance among those different roles? I saw, from your introduction, that the role as creator is the most important one, could you please tell us the reason? ACS (ACS hereinafter referred to as A)

A: 您是一名设计师、人文主义者、文化人类学家、商人以及创造者，那么您是怎么在这些身份之间保持平衡的呢？我从您的简介上看到，创造者的身份最重要，能给我们详述一下原因吗？ACS（ACS以下简称A）

R: It seems to me that keeping the balance between all the roles that you mentioned is possible only because the occupation which is my profession, and at the same time my passion, is a multiple task that requires being active in all these areas. Being a designer means being in all of these roles simultaneously. Participation in culture, understanding of the world around me, being a philosopher focused on reality, and being at the same time really efficient in designing is the quintessence of how I understand my profession, of how I approach the tasks that are formulated for me in the field of design. In this way I understand the responsibility of design which I offer to people and to my customers. (Robert Majkut hereinafter referred to as R)

R: 我想对我而言，能够在这些身份之间保持平衡仅是因为我所从事的职业同时也是我的热情所在。它是一个多方面的事业，需要我在这些方面保持活跃，因此作为一名设计师也就意味着同时拥有这些角色。融入文化，了解我身边的世界，成为一个关注实际的哲人，同时在设计方面充满效率是我对我的职业的理解，也是我在设计领域处理制定项目的方法。通过这种办法便能很好地明确我对人们和我的客户的责任。（Robert Majkut以下简称R）

A: Robert Majkut Design studio has been present on the market for 16 years, and your projects were described in media as pioneer and breakthrough, trend-setting. Are you satisfied with your achievements so far? ACS

A: 罗伯特·迈酷设计工作室成立至今已有16年，您的作品被媒体描述为先锋、突破以及引领潮流的作品。您对它迄今为止的成就满意吗？ACS

ING Corporate Banking
ING 企业银行

Design Agency | Robert Majkut Design

The basis of the concept was the idea of creating new, abstract space for corporate customers sector, which is extremely important for the bank.

Private banking, as a specific way of building bonds between the bank and the wealthy client, requires special approach in terms of design.

The proposed solution is dedicated to corporate customers, who are a fusion of retail and VIP customers. However, the scale of turnover, position of the customers and the character of meetings, place these clients in a more privileged sector, with higher requirements concerning the standard of service and more sensitive to the way of building relations and environment in which it takes place.

Taking into consideration functionality, solutions used concentrated on maximum effectiveness and perfect matching to the character and duration of meetings, required technological equipment and the way of their organization. Central, spacious, representative reception is also adjusted for the waiting room with its business TV channels, coffee corner for people who are waiting and also for those who are in the cabinets, and the communication network which enables easy access to all cabinets. Electronic system, controlled form the reception desk, shows occupied and free cabinets by displaying particular symbols near doors. Cabinets – adjusted to bigger meetings – are equipped with standard infrastructure elements.

The client zone is linked directly with the back office with the whole working team, offering the customer service. In the back office there are also cabinets of the management board and directors of particular departments and regions. Apart from the furnishing, the standard of the back office does not differ from the customer service zone, and in fact is the simplified extension of it.

企业客户区对银行而言极其重要，整个设计概念的核心在于将这一区域打造成一个新颖、抽象的空间。

私人银行作为联系银行和富裕客户之间的特有渠道，在设计上需要更为特别的处理方法。

初步方案关注于企业客户，其中包括个人投资者和VIP客户。然而，依据资产、地位的不同和会面的特点，这些客户能够享有更多的特权，因此他们会对服务提出更高要求并对建筑与他们所处的环境更为敏感。

考虑到功能性，本案所采用的方案在最大程度上关注其有效性以及能够完美地满足不同特点和不同时长的会面、所需要的技术装备以及他们的组织方式。最为中心、宽敞而具代表性的接待处也依据等候室进行了相应调整，为等待的客户和那些正在办理业务的客户提供了商业电视频道、咖啡角以及保证客户更加方便地进入业务隔间的通信网络。由接待台控制的电子系统通过位于门边的特别标识来显示使用中和空闲的业务隔间。为了适应更大型的会面，隔间中配备有标准会议基础设施。

客户区直接与后勤办公部门相连，有一整个工作团队随时待命为他们服务。后勤部门中同样有一些供董事会和一些特殊部门或区域领导使用的隔间。这一部门的标准丝毫不低于客户服务区，它就像是客户服务区的一个简单延展。

R: In my work I try to combine my real two passions: a passion for design and a passion for philosophizing on the future, on how the world can and should be changed to be better than the one we live in today, on what actions to take to change what pinches me and disturbs me in it, on what I think should be changed, because this change for the better seems to be obvious. Perhaps our projects are often seen as pioneering, because of this attitude towards the future. In our business we always try to start from what we found, treat the common solutions as a starting point and move on, looking for solutions that make sense not only today but also tomorrow. We design to solve our future problems, introduce new values changing our environment for the better.

R：我总是试图在我的设计中融入我的两项热情所在：一是对设计的热情，一是对未来哲思的热情。探究世界应如何变得比我们现今居住得更好，探究我们该如何行动以改变摆在我们面前的阻碍，以及探究那些我认为需要改变的东西，因为想要变得更好的愿望是如此的显而易见。也许是因为这种对未来探究的态度，我们的作品常常看起来很前卫。对于我们的事业，我们总是希望由我们所能发现的开始着手，将一些普遍办法作为出发点并推进，探索一种不仅适用于现在、也适用于将来的办法。我们的设计旨在解决未来的问题，引入新价值来使我们的环境变得更好。

A: Your motherland, Poland, is a country with rich atmosphere in art & design, for instance, Polish poster, with its unique design style, represents its high artistic standard and gets high reputation around the world, which is far beyond other countries. Is there any influence on shaping your design style? I mean the Poland's artistic atmosphere. ACS

A：您的祖国波兰是一个艺术设计气息非常浓郁的国家，例如，波兰的海报，其设计风格独特，体现着高超的艺术水准，在世界享有盛誉，是其他国家望尘莫及的。这种艺术气息对于您个人的设计风格的形成有什么影响吗？ACS

R: What you ask about is a very complex issue. For example, the Polish poster is something which brought up all creators in Poland, all are very familiar with this style. Certainly the impact of the Polish poster school on the contemporary Polish graphic design is very prominent. However, it seems to me a little less important and less influential on the other creative disciplines, such as new media, interior design and architecture in general, the reason for this is that the poster communicates itself in a different medium, uses other means of expression. Nevertheless, we cannot negate the fact that in Poland we grow up in some common aesthetic, which is easily identifiable in the Polish poster, and in this respect certainly the roots of the creators of many disciplines of art are in Poland somewhat common. I also use the benefits of this experience and this knowledge in my thinking about design. As for me, Polish poster is primarily a concept for synthesizing thoughts and looking for the most adequate and precise means for what is to be said. And at this level is very close to me.

R：你所问的是一个非常复杂的问题。例如，波兰海报就像一个哺育波兰创作者成长的存在，所有人都对它的形式非常熟悉。当然，波兰海报学院对当代平面设计的影响是十分显著的。然而，它对我而言却不如新媒体、室内设计和建筑等领域那么重要或是有影响力。其原因在于海报交流本身通过一种不同的介质，使用其他的表达方式。但我们也无法忽略它的影响力：在波兰，我们成长于共通的审美中，这些在波兰海报中非常明显。从这个方面来看，确实，波兰许多领域的创作者的根基或多或少都是相同的。在设计过程中，我也同样得益于这些经历和知识。于我而言，波兰海报首先是一个合成思想和寻找最合适、最精确处理方式的概念代表。在这个层次，它与我紧密相连。

A: In Poland, you are regarded as the idol of interior design; In abroad, you are regarded as an excellent designer because you changed Polish public space and promoted the quality of Polish life. According to your experience, what is the most prominent feature of Poland design style? ACS

A：在波兰国内，您被称为内饰设计的偶像；在国外，您被认为是一个改变波兰公共空间，提升了波兰人生活质量的优秀设计师。依您的经验来看，波兰设计风格最突出的特征是什么？ACS

R: Generally speaking, in the era in which we live it is more and more difficult to talk about the "nationality" of design. Yes, there are some crystallized distinctions, some qualities and features that developed over the decades in certain cultural surrounding, but together with the technologic changes of recent times and in a very dynamic global exchange of goods and information, all national styles started to affect each other deeper than ever, interlace with each other, percolate… and it is very hard to pick out a specific characteristics of national designing anywhere. Of course there are such centers such as the U.S.A., where

Kronverk Cinema
Kronverk 电影院

Design Agency | Robert Majkut Design

Kronverk Cinema is a leading cinema network on the Russian market. It encompasses 15 cinema theaters in major cities in Russia (including 8 in St. Petersburg) and 1 in Ukraine.

The concept developed by Robert Majkut Design was realized in one of the cinemas in Moscow. It allows to be repeated both in new and existing cinemas as a coherent and consistent solution of the spatial organization and visual identification.

Recognizable elements of Kronverk's brand identity – the distinctive logo and colors – were the starting point for this project.

The concept is realized as a smooth development of the logotype's shapes and its color balance. All spatial geometric forms present in this project are the consequence of a certain order of lines, which comes from the trademark – most of all from the distinctive symbol of crown. Derivation of these lines, their multiplication and crossing has created a flexible array of forms, which can be modified into various modules. In this way for each wall or ceiling there is a common matrix, one can find their counterpart in the elements of the logo.

Kronverk brand's underlying colors were repeated in each zone, altered through different styles and moods. The choice of materials and finishing techniques was important in this project – they had to be both aesthetic and economic, repeatable in different configurations – as it is the optimal solution for network projects. The decorative function in the interior, however, is played not only by the materials, but also by the forms and colors, and above all by the interesting and surprising idea to combine the seemingly distant aesthetics.

This design in a modern way refers to the Russian decorativeness.

Kronverk 电影院是俄罗斯市场上首屈一指的连锁电影院，其中15个分布于俄罗斯各主要城市（8个位于圣彼得堡），一个位于乌克兰。

其中，位于莫斯科的电影院由罗伯特·迈酷设计工作室进行设计。这个电影院能够重现一些新开的或是已有的电影院元素，延续了对空间组织和视觉识别的处理方式。

Kronverk 的品牌形象识别元素——那别具特色的商标和颜色——是整个设计的出发点。

设计概念的实现来自于对商标形态和色彩平衡的顺利开发。这个项目中所有的空间几何形态都按照一种特定的线条顺序——这种顺序取决于它的商标——大部分来自于它与众不同的皇冠标志。以这些线条为源，通过层叠和交叉，创造出一种灵活排列的形态，这些形态能够进一步衍生出不同的模式。每个墙面和天花上都有着相同的矩阵，人们能够在商标元素中找到它们的源头。

Kronverk 品牌的底层颜色被重复应用于各个区域，并根据不同的风格和基调进行修改。材料的选择和粉饰技巧在这个项目中尤为重要——它们需要在充满美感的同时兼具经济性，能够在不同的结构配置中进行重复使用——这是对于系列项目而言最佳的选择。室内装饰效果并不全由材料来体现，还包括形式和色彩，尤其是运用一些有趣而令人惊喜的理念将看似遥不可及的美学结合进装饰中。

整个设计参考了俄罗斯装饰风格并按照一种现代的手法进行实现。

the local traditions are still very strong, for instance of the Wild West or even Pop Art, which specificities are easy to capture and their influence can be seen up to this day. In such a "young" and inhomogeneous design as Polish, such phenomena are still shaping, emerging, because some time has to pass before the accumulation of tradition is so strong that someone will be able to formulate a characteristic of the Polish style.

R: 总体而言，在现今我们生活的时代谈设计的"国籍"愈发困难。是的，设计确实存在着具体的不同，包括一些在近几十年发展起来的、处于一个特定文化环境中的品质和特点。但是随着近年科技的变化和愈见活跃的全球商品和信息的交换，所有的国家特色开始出现从未有过的深度相互影响，彼此互相穿插渗透……因此要想拣出一个特定的本土设计特色是非常困难的。当然也会有这样一些地方，例如，美国，本土传统还十分浓厚，例如狂野西部，甚至波普艺术，它们的特色易于掌握，它们的影响力在今天依然可见。在波兰这样一个设计年轻而不均衡的国家中，这样的现象正在逐渐成型、显现，因为传统的累积需要时间，这样人们才能真正表达出所谓的波兰特色。

A: From one design comment, I saw that you are sensitive to humanoid forms, oblong, round, close to nature, and you pay special attention to recording of the human gesture, so in your projects you are always inspired by human movement and the man himself. Do you agree with this comment? Would you please briefly interpret this? ACS

A: 我看到一则设计评论说您对于人形、长方形、圆形及亲近自然等因素很敏感，您特别重视人体姿势的刻画，所以您的设计总是被人类活动和人类自身所启发。您是否认同这一评论，可以为我们简要解释一下吗？ACS

R: I fully agree with that comment! But I think that when it comes to sources of inspiration, speaking about nature as a common denominator is a very big generalization. Because under the word "nature" there is a fish, a stone and a tree. The wealth of forms is endless, inexhaustible. If I think about myself, about what is for me the most important in this giant resource of inspiration, I observe that I focus the most on those aspects that are closely connected to the relationship between human and the environment, the architecture. I'm interested in how a man in his presence enters into a certain relation with architecture, and as such a man is a reference point for me to learn how this relation is built and, what is very important, how the architecture surrounding a man understands him in the cultural, physical and biological sense. Such aspects as the human condition, man's motor skills, abilities, predispositions, including those that are implied by culture, are very important to me. This understanding (or misunderstanding) of human being is often manifested in very simple phenomena, in the functionality of design, such as the height of a step of stairs or the depth of chairs that we sit on, but also on a much larger scale, in what we feel as a kind of "personality" of space: warm, lofty, aggressive or soaring. Both in the micro and macro scale, through manipulating with specific measures, design affects the relationships in the human world, and this is for me most fascinating.

R: 我完全同意这一评论。但我认为当谈及设计灵感来源时，将自然作为基数是一种非常笼统的概括。在自然项下可以有鱼，有石头，有树木。形态的丰富多彩是广阔无边、不可穷尽的。当我思考对我最重要、最巨大的灵感来源时，我发现我总是特别关注那些和人与环境密切相关的方面——建筑。我对于人们如何以自身存在进入到一个与建筑发生特定关系的空间，同时，如何将这个人作为建设这一关系的参考点，以及很重要的一点，围绕这个人的建筑环境如何在文化、物理、生物感官上与其相契合，非常感兴趣。人的基本情况包括运动技能、能力、素质，包括那些受文化影响的部分，对我而言都非常重要。对于人类的了解或是误解常常显现在一些非常简单的现象中，在设计的功能性，例如，楼梯阶级的高度和座椅的深度，当然还包括更广的范围，即能够让我们在空间中感受到一种"个性"：温暖的、高贵的、奋进的或是无拘无束的。无论在微观还是宏观上，通过采取一些特定的手段，设计能够影响人类世界的关系，而这点对我来说是最为让人心驰神往。

A: You have designed many wonderful projects, then which one is your most memorable project? Would you please share with us the story behind it? ACS

A: 您设计了很多极好的项目，令您最难忘的一个设计项目是哪个？可以跟我们讲述一下它背后的故事吗？ACS

R: It is always difficult for me to answer such questions, because of the many projects that we have completed, each has a story behind it that is worth telling. Each of them also has a story to tell, an inner story expressed in means of design. I try not to valuate these stories, not to compare them, because these are tales with different characters, with their own narrative and with their own presented world, so it doesn't really make sense for me to compare them. I simply like the most of my projects, for very different things. I

Multikino Złote Tarasy
Multikino Złote Tarasy 影剧院

Design Agency | Robert Majkut Design

It is without doubt one of the most recognizable interiors of Poland, a spectacular and unique place, both in terms of design and technical solutions. The cinema is located in the busiest and most energetic parts of Warsaw, in Złote Tarasy commercial center. Its owner, Multikino brand, being part of ITI Group, set up a real challenge – to create a focal point for cinema world and a place of amusement that would reach the level previously unknown in this part of Europe.

The very stylish, elegant and chic interior is a distinctive feature of the facility. Three cinema levels provide functionality which will meet the needs of various groups of cinema-goers, as Robert explains: "The design is an attempt to create such a space, and to look for suitable aesthetics for such a facility. By upgrading the standard, alluding to the celebrations of the film world, by placing innovativeness, uniqueness and nobility in the focus, I tried to re-introduce nobility to what had been impoverished – combine the convenience of a multiplex with the atmosphere and dignity that we know from cosy cinemas or grand theatres and concert halls. No one builds such cinemas now, but I truly believe that it makes a lot of sense, as we again want to live in a better reality, and of a very good quality." It is a mix of two worlds, a crossing between future and past. Modern forms with decorative motives known from other eras, such as renaissance and baroque. Vibrant colors, spectacular lighting and top quality materials provide a cinematic experience of an unparalleled standard.

The project of Multikino Złote Tarasy set a new, high standard in the Polish entertainment market, as a result of brave and unconventional approach to places of mass culture.

毫无疑问，就设计和技术方案而言，这个位于华沙最为繁忙和充满活力的地区——Złote Tarasy 商业中心的电影院项目是波兰最具辨识度的室内设计以及最壮观独特的场所所在。它的所有者，Multikino 品牌，作为 ITI 集团的一份子，面对的是一次真正的挑战，即打造出一个影剧院世界的聚焦点，一个能达到欧洲这一领域中全新水平的娱乐场所。

这个十分时尚、优雅、独特的室内设计成为了整个电影院的显著特色。三层的影剧院保证了其功能性，很好地满足了不同观影人群的需要。正如罗伯特解释道："这个设计试图创造出一个空间，并同时赋予其合适的美感。通过完善标准，含蓄地向整个电影世界传达祝贺，以及关注创新性、独特性和高贵品质，我试图将我们从诸如舒适电影院或者大剧院和音乐厅中感知到的氛围和高贵与综合电影院的便利相结合。如今还没有人尝试过这个类型的电影院，但我坚信这非常有意义，因为我们总是渴望生活在一个更好的世界，拥有更好的品质。"这是两种世界的混合，穿插着过去与未来。现代的设计风格中融入了其他时代的装饰特色，例如，文艺复兴时期风格和巴洛克风格。明快的色彩、引人入胜的灯光以及顶级材料的使用必将带给人们一次无与伦比的观影体验。

对大众文化场所的这次勇敢的打破常规的尝试使得 Multikino Złote Tarasy 影剧院项目在波兰娱乐市场中树立了一个新的高标准。

appreciate some for their scale, pageantry, the other I like for the fact that I was able to realize some of my ideas that was in my head for longer time, again others I appreciate for their complexity, their uniqueness or innovativeness which they introduced in their own category. However, I think one of my most beautiful designs, which is a beautiful story from the beginning to the end, is Whaletone, a piano, which I designed inspired by an encounter with whales.

R: 这个问题总是很难回答，因为在我们完成的每个项目背后都一个值得讲述的故事。这些故事表达了我们的设计方式。我试着不去评价或是对比这些故事，因为它们就像童话故事般有着各自不同的角色、各自的讲述者和各自所呈现的世界，因此我觉得将它们进行对比并不十分有意义。我很单纯地喜爱我的大部分作品，爱着不尽相同的东西。或是它的规格、它的华丽，或是它实现了我那种萦绕许久的想法，亦或是它们各自方案中的繁复、特别、创新。然而，我最美的设计之一是《Whaletone》，一架钢琴，它的故事从开始到结束都是那么美，其设计灵感来自于我遇见的鲸鱼。

A: With wide design field, your design covers interior design, industrial design and Corporate Identity. Moreover, you always keep good balance among them. Do you have any unique approach or effect experience to maintain your own creativity in different design fields? ACS

A: 您的设计领域很广，包括了室内设计、工业设计以及企业形象设计，并且您面与面之间都平衡得很好，您有什么独特的方法和有效的经验保持自己在不同设计领域的创造性吗？ ACS

R: The complexity of the tasks which we deal with stems from our experiences, from the more and more complex projects which in the process of evolution of our practice we decided to face. This shaped our know-how very extensively. Now we're at that point of development, in which this diversity of experiences is paying off, because they were collected from such different fields. We manage to combine them into a

Orange Cinemas
橙电影院

Design Agency | Robert Majkut Design

The interiors of Orange Cinemas consist of Piano Bar zone, private VIP-rooms, Cigar Club, and three luxurious cameral auditoriums: Black Room, Orange Garden and Pink Sky. Lobby is a unique cameral club space with the collection of the latest works of art.

Robert Majkut, with his great experience, has created unique cinema interiors for many years. Also this project has a unique character. The designer tells a story about moving, turning forms, movement, impressions, and illusion of the cinema. All these elements are expressed by colors, textures, and light. The space is created thanks to variable light of big screens, sounds from Piano Bar, and the intense play of colors.

Logo, inspired by manually painted Chinese symbols, was the basis for the design of Orange Cinemas, interior. The shape of the bar, carpeting patterns, shapes of door and table holders, openwork curtains, quilts, and wall patterns are the transformation of the sign form.

Smooth surfaces are the continuous game of escaping and overlapping colorful spots – starting with orange, through pink and dark chocolate hues, to dominating black. Big contrasts are on the one hand a background for multimedia performances, but on the other hand they are a perfect place for intimate meetings in the middle of the film scenes.

All furnishing elements – sofas, tables, bar stools and lighting – were individually designed by Robert Majkut. All is kept in consequent black, orange and pink coloring.

Material used in the concept are shiny stone, mahogany veneer, silk wallpapers, wool carpeting, leather, felt, and polished steel.

橙电影院的内部包括一个钢琴吧台区、VIP专属区、雪茄俱乐部和三个豪华观影礼堂：黑色空间、橙色花园和粉色天空。大厅是一个独特的观影俱乐部空间，有着最新的艺术收藏作品。

罗伯特·迈酷凭借自己丰富的经验，设计出了许多独特的电影院内部空间。这个项目同样也有着自己独有的特点。设计师好似在讲述一个故事，移动、转换着电影院的形式、动作、印象和幻像。所有这些元素都通过色彩、纹理和光线来进行表达。多亏了来自巨大屏幕上不断变化的光线、钢琴吧的声响和色彩的集中展现，才使得这个空间得以实现。

Logo，得灵感于手工绘制的中国符号，是整个橙电影院室内设计的基础。吧台的形状、地毯的花纹、门把手和桌垫的形状、透空式窗帘、靠枕和墙壁的图案都是这一标志的变体。

平滑的表面是彩色光斑们一场永不停歇的逃离又相互重叠的游戏——先是橙色，渐渐变为粉色再到深咖色，以此主导着黑色空间。一方面是可用于多媒体表演的大背景；而另一方面，电影场景的中部是私人会谈的绝佳场所，这二者形成了鲜明的对比。

所有的家具——沙发、桌台、吧台高脚凳和灯光——都出自罗伯特·迈酷的设计，均以黑、橙、粉作为系列色。

设计概念中运用的材料有光滑的石头、桃花心木薄片、丝质墙纸、羊毛地毯、皮革、毡制品和抛光钢材。

single comprehensive service, a consistent proposition for the customer, so we can have the maximum impact on the results and the quality of work, we can act in a wide angle controlling all aspects of the tasks that are entrusted to us. It is a huge responsibility, but also creatively a very comfortable situation, very uplifting, because it allows us to operate with all the measures which are necessary for the project to develop.

R: 我们所接项目的复杂性来自我们的经验，来自于那些在越来越复杂的项目中我们不断想要进行完善的处理办法。这一点使得我们的专门技术范围非常广阔。如今，我们处在发展的某个点上，在这个点上我们因为涉及如此多样领域和由此获得的丰富经验将逐步得到回报。我们试着将它们与单一的综合服务相结合，为客户提供连续的建议，以便我们能够在最终的结果和项目品质上发挥最大的影响力并在大角度上控制我们所负责的项目的各个方面。这是一项巨大的责任，但同时因为充满创造性而令人倍感舒适，十分让人振奋，因为它让我们能够参与进项目发展的整个过程中。

A: It is said that you have high requirements toward brand and investors, is it real? As an international design studio, how do you choose your suitable clients? ACS

A: 据说您对品牌和投资者的要求很高，是真的吗？作为一家国际设计公司，您如何挑选适合自己的客户？ ACS

R: Our commitment to high standards, which is the value of our design company, is really acting in the interest of the project, which means also in the interest of the investor. I believe that if we realize a

project, we (I say "we" in the sense of all participants of the whole process) have to do it right, the best we can and the best as we know that this should be done. In our design work we should always be at the upper limit of our abilities, in order to make the effect be exactly what we planned. So that the tasks that are put before us, sometimes difficult and demanding, always find their best solution, so that the quality of what we create was high, and the effectiveness of our work was high as well. I intend to never let go, which is often a natural tendency of many of us during lengthy and tedious processes – to turn a blind eye to some imperfections that result from some or other difficulties. I always make many efforts that the whole process in all its stages is kept in the highest standard, despite the fact that it is not easy, because after all, everyone is later satisfied with that.

R: 高水准是我们公司的核心价值。我们对高水准的承诺表现在确实地以项目的利益而行动，同时意味着为了客户的利益而行动。我相信当我们实现一个项目，我们（我说我们是指项目全过程的所有参与者）必须做我们应该做的，做到最好，取得我们所想要取得的最好成果。在设计工作中，我们必须总是处于我们能力的上限，准确实现我们的预期效果。因此我们面对的项目，有时候也许十分困难，具有挑战性，但我们总能找到最好的解决方案，保证整个项目的高品质和高效率。我们中的许多人在冗长繁琐的处理过程中很自然地倾向于对一些困难产生的不合标准视而不见，但我总是不愿妥协。我总是努力将项目的全过程、每个阶段保持在高水准，尽管这么做并不容易，因为毕竟，每个人都是在后来才表示满意。

A: Is Orange Cinemas, located in Beijing, your first project in China? What is your future direction in China? ACS

A: 北京的橙电影院是您在中国的第一个项目吗？今后您在中国的发展方向是什么？ ACS

R: Orange Cinemas is the first project implemented in China, but it is not the very first project that we have done for a Chinese customer. The project, which chronologically preceded the implementation has a much larger scale, and unfortunately is all the time waiting for the opportunity to be implemented, to which we really look forward. I believe that China is a very interesting, fascinating market for design. Our proposal, which was the Orange Cinemas in Beijing, has been very well received. That is a signal for us that we are able to propose much more. So I hope that this is not our last, but the first of many implementations in China.

R: 橙电影院确实是在中国完成的第一个项目，但它并不是我们为中国客户做的第一个项目。在此之前的项目有着更大的规模，但遗憾的是我们还在等待和期待最后完工的机会。我相信中国有着非常有趣和令人着迷的设计市场。我们的项目——北京橙电影院获得了非常不错的反响。这给了我们能够在这个市场上走得更远的一个信号。因此，我们希望这个项目能成为今后众多中国项目中的第一个而不是最后一个。

A: I knew that you designed a piano with a cool whale shape. Are you a piano lover? Except for design, what other hobbies do you have? ACS

A: 我知道您设计了一款超酷的鲸鱼外形的钢琴，您是一个钢琴爱好者吗？除了设计之外，您还有哪些兴趣爱好？ ACS

R: Unfortunately, I do not play the piano, but I really like music. It is just amazing that I'm not a musician, I'm not an instrumentalist, and one of the most amazing adventures that happened in my life so far was to design a piano. This is a project that I simply dreamed of and I moved it with remarkable consequence into the reality. I appreciate the natural sound of the piano very much, but what we did and we want to do in Whaletone piano is an escape to the front, it is creating a new form for a new class of instruments, similar to what happened once when electric guitar appeared, growing out from the acoustic guitar. Whaletone seems to me a new voice in thinking about such type of projects, not just the next step in the formal evolution of piano. I have to admit, it's a phenomenal experience to work on such a distinguished platform, which such a serious instrument definitely is. Of course, in addition to design, there are other things and events in my life that I'm very passionate about, but the most important is my son.

R: 很遗憾，我并不弹钢琴，但我热爱音乐。我不是一个音乐家，不是一个乐器手，在我的生命中发生的最令人惊讶的事是我设计了一架钢琴！这是一个我仅仅梦想过的项目，我试着将一种非凡的理念转变为现实。我非常喜爱钢琴自然的声音，但我所做的和我希望这架鲸鱼钢琴所呈现出的是一种打破常规的形态，一个全新的乐器类别，类似于由木吉他演变而生的电吉他的首次问世。鲸鱼钢琴对我而言就像引领我思索物体类型的一道新声音，而不仅仅是常规钢琴的演变。我必须承认，在一个如此特别的平台上进行设计真是一

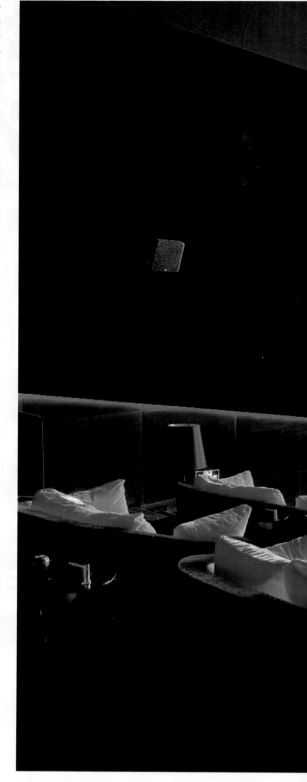

场非凡的经历。当然,除了设计之外,我对生活中的许多方面充满热情,但最重要的是我的儿子。

A: Could you please share with us your future plans or your new project? ACS

A: 可否跟我们分享一下您未来的一些计划或者您的新项目? ACS

R: I think that what I'm currently working on will soon, in the nearest future, be coming to see. In fact, all my work is the work of the future, because what I do, what is on my table right now, is the future for the others. My plans are very ambitious and I would like to realize a lot of projects which appear on my horizon, in many places in the world. I hope that at least some of them will be implemented within the next few years.

R: 我想我现在着手在做的事在不久的将来就能看到成果。实际上,我所有的工作都是为未来而做,因为我现在在做的和那些摆在我桌面上的事,正是其他人的将来。我的计划充满了雄心,我想要在这个世界上的多个地方实现多个项目。我希望在未来几年中至少能够完成它们中的一些。

Whaletone
鲸鱼钢琴

Design Agency | Robert Majkut Design

Whaletone is a professional musical instrument for individualists. Its beautiful two-color, black and white case of the classic model, resembles a whale which breaks the waves, visibly arching its body.

Fully bespoke elements, like the color of the case or upholstered "soul" of the piano are in the standard offer. Its true uniqueness is reflected in many variables. Each and every unit is always made on individual request, dedicated to its owner, adjusted to their preferences and requirements, marked with a personal number.

The beautiful, modern shape is not the only virtue of Whaletone. This fully professional stage instrument is able to respond to many music challenges. That is why it is perfect for connoisseurs who look for objects combining many aspects in one exquisite frame.

The heart of the instrument is one of the greatest stage pianos, with a Super Natural Piano technology, ensuring the highest quality of the classic grand piano sound, enriched with the elements crucial for professional musicians. It is equipped with a comfortable keyboard, perfectly transmitting the dynamism of sounds, and an Ivory Feel surface, closely imitating the real ivory and ebony keys. The sound is strengthened with a four-channel amplifier and is emitted from the loudspeakers, creating a precise, dynamic sound scene.

Due to its form and possibilities, Whaletone will perfectly fit in on a professional stage, as well as in a private room, in a club, hotel or on a yacht. Every detail of the instrument has been precisely crafted. Every element has been put through multiple tests and countless improvements, with the aim of creating a perfect piece of design.

鲸鱼钢琴是一件个性化定制的专业乐器。漂亮而经典的黑白按键象征着鲸鱼破浪翻腾的身影。

按键的颜色和钢琴的灵魂——它的内部配置等全定制元素都为统一标准。它的独特性反映在许许多多的外在变化上。每一架钢琴都按照个人要求进行设计，依据主人的偏爱和要求并标上个人编号，使其完全成为其主人的归属物。

优美而充满现代感的造型并不是鲸鱼钢琴的唯一看点。这个极具专业性的舞台乐器能够应对许多音乐挑战。这就是为什么对于那些既追求乐器各方面专业性、又追求精致轮廓的行家而言，它能够成为一个完美的选择。

这件乐器的核心是一架极出色的舞台钢琴，应用有 Super Natural 钢琴技术，保证了最高品质的古典钢琴音，以此强化了对于专业音乐家而言至关重要的那些元素。它装备有舒适的键盘，能够完美地演绎音乐的美妙悦动；象牙般滑腻的表面，一如真正的象牙和乌木。声音借助一个四通道扩音装置得以加强，通过扬声器流泻而出，创造出一种精准而动态的音乐场景。

鲸鱼钢琴依凭着它的形态和可能性潜力，不仅适用于专业舞台，也同样适用于私人处所、俱乐部、酒店或是游艇。乐器的每个细节都经过精心处理，每个元素都经过多次测试和无数次改善，力求创造出一个真正完美的设计作品。

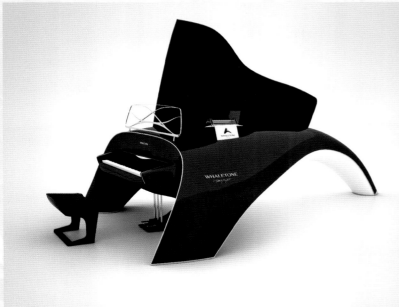

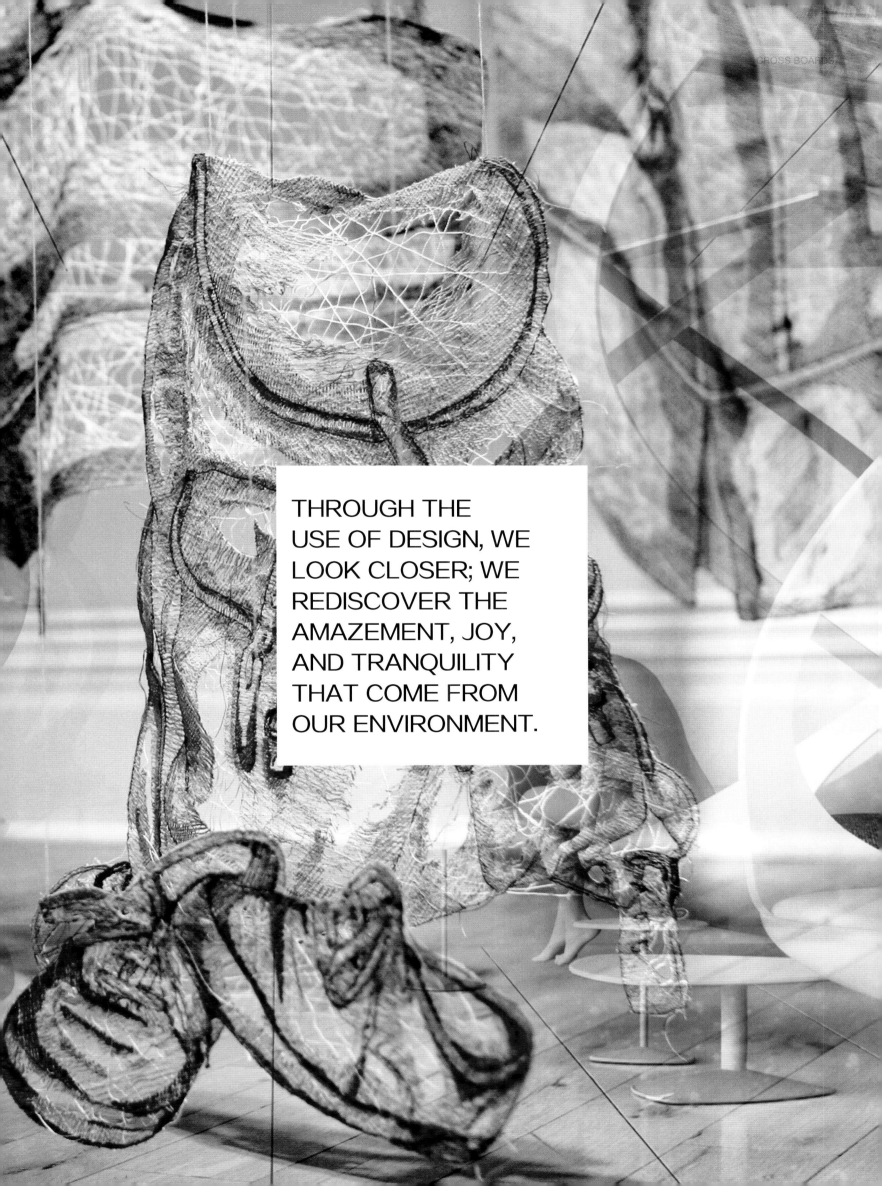

THROUGH THE USE OF DESIGN, WE LOOK CLOSER; WE REDISCOVER THE AMAZEMENT, JOY, AND TRANQUILITY THAT COME FROM OUR ENVIRONMENT.

CROSS BOARDS | ARTPOWER CREATIVE SPACE

Thread/Machine
Embroidery Articles 线绣摆件

Designer | Amanda McCavour

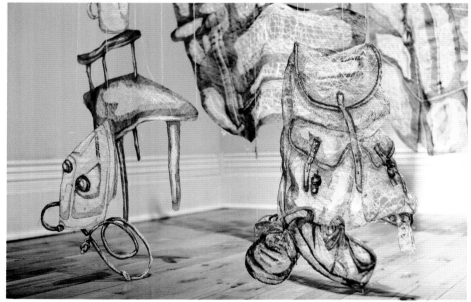
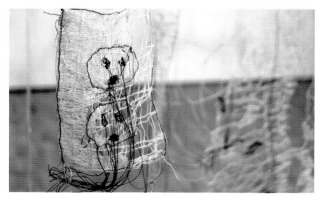

Amanda McCavour is interested in thread's assumed vulnerability, its ability to unravel, and its strength when it is sewn together. Here includes several of her projects which are made of thread with the help of machine embroidery:

Stand-In for Home: This installation is a thread rendering based on part of Amanda McCavour's kitchen in her previous home.

* produced with the support of the City of Toronto through the Toronto Arts Council

Superspiroscribble: This piece is a drawing made out of thread, an accumulation of lines that creates a gradation of color.

*produced with the support of the Ontario Arts Council

Living Room: This piece based on her old living room in her old apartment. The thread drawings act as tribute to a room that once was.

*produced with the support of the Ontario Arts Council

Ice Box: In the past, both glass shards and cotton batting have been used to create "crafty" versions of artificial snow for Christmas trees. Amanda McCavour made frost and ice crystal forms out of cotton thread.

*produced with the support of the Ontario Arts Council

Floating Garden: "Floating Garden" addresses the history of botanical themes in stock embroidery, taking flowers out of the context of embroidery "kits" and moving these images into an installation to create an experiential environment.

*Produced with the support of the Ontario Arts Council, The Surface Design Association and la Maison des métiers d'art de Québec

Amanda McCavour 对用线营造一种柔弱感非常感兴趣，它可以被解开，但当缝在一起时又具有自身的力量。这里展示了她的一些线制的作品，由机绣辅助制成。

家居摆件：这个装置是利用线模拟家居摆件，模拟的对象为 Amanda McCavour 原先家里的厨房。

* 制作过程中得到了多伦多市艺术委员会的支持。

挂件：这是一个由线制成的绘画作品，通过线的积累来创造不同色彩层次。

* 制作过程中得到了安大略艺术委员会的支持。

起居室：这个作品以她原先旧公寓的起居室为基础，表达了她对那些曾经拥有过的房间的敬意。

* 制作过程中得到了安大略艺术委员会的支持。

冰盒：过去，玻璃碎片和棉絮被用来制作圣诞树上的人造雪，而 Amanda McCavour 用棉线来营造雾和冰晶的效果。

* 制作过程中得到了安大略艺术委员会的支持。

飘浮花园：飘浮花园强调了刺绣中以植物为主题的历史，它将花从刺绣背景中提取出来并把这些意象放进这个项目中，以创造一种可感的环境。

* 制作过程中得到了安大略艺术委员会、平面设计协会和 la Maison des métiers d'art de Québec 的支持。

Nexuscouch is a conceptual project made on the fifth edition of the "Vision of the leisure 2012: Missing element" organized by the Polish furniture manufacturer KLER S.A.

The theme of the competition concerned the product design and its place in the collection of the commercial manufacturer. The organisers require from designers a critical evaluation of the deals and enhanced with an important element, which will raise the value of the collection.

The main inspiration for this project was the missing link in the chain of DNA, as a metaphor for a missing item in a collection of furniture KLER.

Nexuscouch is in terms of functional furniture that does not fit in the rigid framework of the generally accepted classification, being the missing link between a sofa and a bed.

Direct reference forms to the DNA chain also has a symbolic dimension, as a metaphor for the individual tastes of each of us.

Another important inspiration was movie, "Blade runner" and its existential questions on the limits of humanity, and this book had a great influence on the final form of the presentation of the project.

"The light that burns twice as bright burns half as long." ("Blade runner")

These are the words that speak of the value and importance of an individual approach. What translates well on the process of product design.

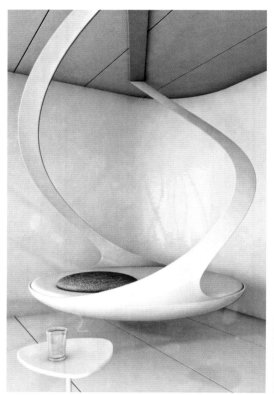

inspiration

尼克瑟斯躺椅是为由波兰家具制造商 KLER S.A. 组织的第五期《休闲视觉 2012：消失的元素》所做的一个概念项目。

这个竞赛的主题围绕产品设计和它在商业制造商中所处的位置。组织方要求设计师在处理及加强重点元素方面有很高的素养，这将提升作品的价值。

这个项目的主要灵感来自于 DNA 链中缺失的节点，作为对 KLER 家具系列中缺少的商品的隐喻。

尼克瑟斯躺椅作为一个功能家具，并不符合普遍接受的分类中的既定框架，而是沙发和床之间缺少的必要的连结。

以 DNA 链为参考还有着一个象征维度，即隐喻了每个人独特的品位。

另一个很重要的灵感来自于电影《银翼杀手》和它对人性的极限提出的问题，这对这个项目最后所呈现的形态有着巨大的影响。

"光的亮度增加一倍，燃烧的亮度就减少二分之一。"（《银翼杀手》）

这句话讲出了对待个人方案的价值和重要性，也对这个产品的设计过程给出了很好的解释。

Nexuscouch

尼克瑟斯躺椅

Designer | be3design

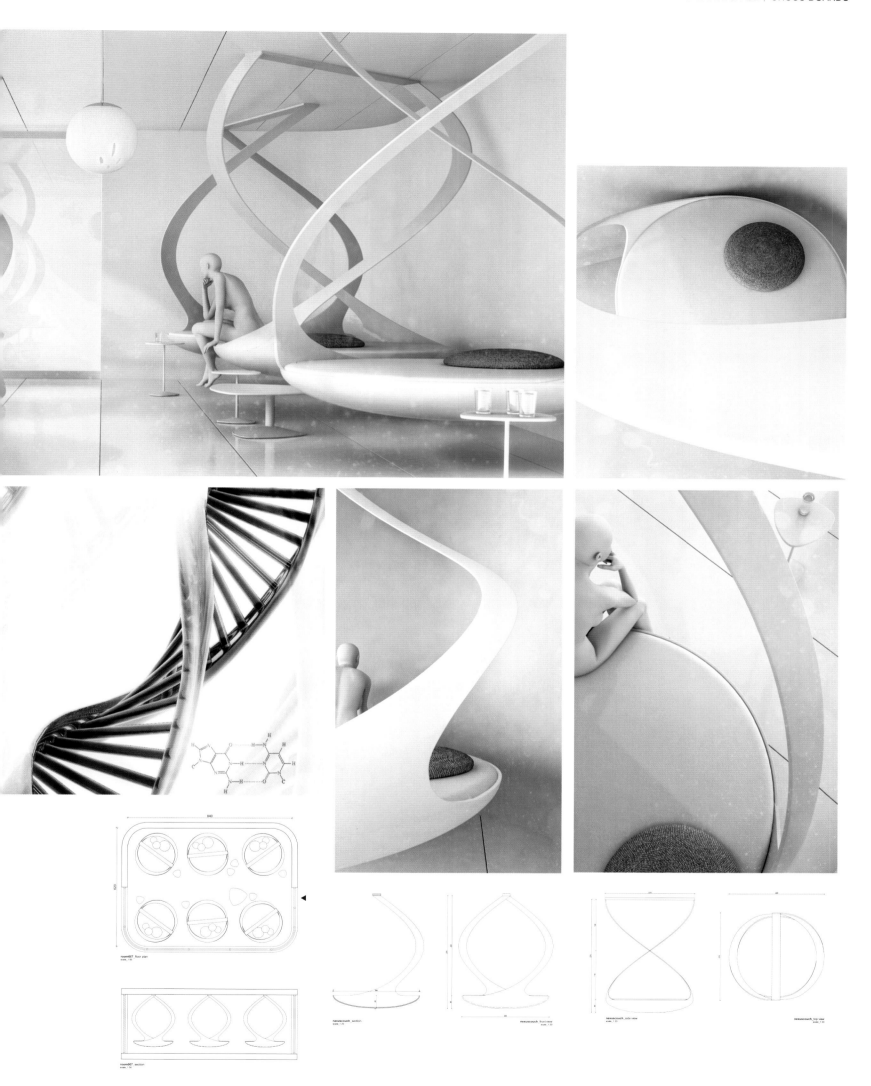

Today, people attempt to subvert impermanence through technology and science. They isolate themselves from the natural world, viewing it from the perspective of a spectator rather than a participant. Going about their daily lives, they rarely notice nor appreciate each unique experience the surroundings offer. For the designer, there is a peace that can be found in even the simplest things a decaying piece of wood, rusted metal, crumbling brick, the growth of moss and lichen. These ordinary elements within the environments offer both visual and physical reminders of our connection with nature. In his observations he also see similarities between the processes that occur in nature and those that drive people. By combining both human and natural elements within his work, he hopes to highlight the fact that we are not separate from nature, but are in fact part of it, and in being so, people are as impermanent as a flash of lightning in the sky.

Through the use of trompe l'œil, people can look closer and rediscover the amazement, joy, and tranquility that come from the environment. At the same time, people witness the impermanence by evenhandedly dialing in on decay. Neither good nor bad, decay is simply a natural process of the world that at times can produce deeply moving and beautiful effects.

Inspiration
from Nature and Man

来自于大自然和人类的灵感

Designer | Christopher David White

如今，人们试图通过科技来颠覆无常。人们将自己与自然世界隔绝，以一个观察者而非参与者的姿态对待自然。在日常生活中，人们很少留意或是赞叹周围环境带来的每个特别体验。对设计师而言，平静甚至能够来自于最简单的事物，例如，腐朽的木头，生锈的铁，碎裂的砖石，苔藓地衣的生长等。这些在人们的环境中稀松平常的事物带来视觉及生理上的提示：人类与自然紧密相连。在设计师的观察中，发生在自然中的现象都具有共通点，而这些点也是他的灵感来源。他在他的作品中将自然与人相结合，以此强调人类与自然不可分离，虽然实际上人类仅是自然的一部分，就像空中的一道闪电。

通过运用错视画，人们能够看得更细致，并重新发现来自环境中的那些惊异，那些欣喜和平静。同时，还可以看到我们的短暂岁月于自然中衰退，不好也不坏，因为衰退仅是我们世界中一个简单的自然过程，它为我们带来的是更深的感动和美好的影响。

Sculpture

Designer | Nuala O'Donovan 雕塑

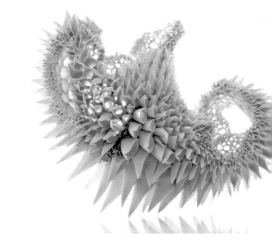

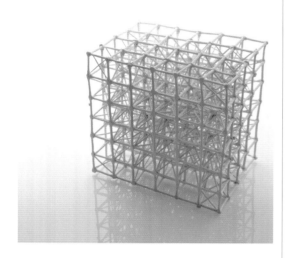

Title of work | Radiolaria, Grid Yellow Center

Material | High-Fired, unglazed & stained porcelain

Technique | Hand-built, Multiple firings

Dimensions | 36 x 24 x 36cm

Photographer | Sylvain Deleu. © Nuala O'Donovan

作品名称：放射性网格黄色中心

材料：高温焙烧无釉彩瓷

工艺：手工制作，多重烧制

尺寸：36cm x 24cm x 36cm

摄影师：Sylvain Deleu. © Nuala O'Donovan

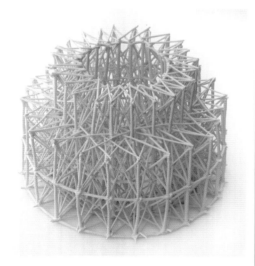

Title of work | Radiolaria, Symmetrical Cylinders

Material | High-Fired, unglazed porcelain

Technique | Hand-built, Multiple firings

Dimensions | 43 x 243 x 27cmH

Photographer | Sylvain Deleu. © Nuala O'Donovan

作品名称：放射性对称圆柱体

材料：高温焙烧素瓷

工艺：手工制作，多重烧制

尺寸：43cm x 243cm x 27cm（高）

摄影师：Sylvain Deleu. © Nuala O'Donovan

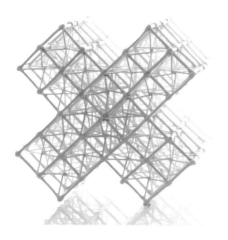

Title of work | Radiolaria, Grid X

Material | High-Fired, unglazed & stained porcelain

Technique | Hand-built, Multiple firings

Dimensions | 36 x 24 x 36cm

Photographer | Sylvain Deleu. © Nuala O'Donovan

作品名称：放射性网格 X

材料：高温焙烧无釉彩瓷

工艺：手工制作，多重烧制

尺寸：36cm x 24cm x 36cm

摄影师：Sylvain Deleu. © Nuala O'Donovan

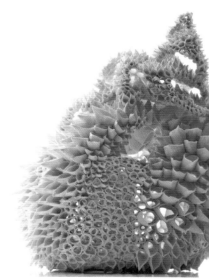

Title of work | Teasel, Grey Fault Line

Material | High-Fired, (1260 centigrade), unglazed & stained porcelain

Technique | Hand-built, Multiple firings

Dimensions | 52cmL x 28cmW x 44cmH

Photographer | Sylvain Deleu. © Nuala O'Donovan

作品名称：起绒草与灰色断层线

材料：高温焙烧（1260度）无釉彩瓷

工艺：手工制作，多重烧制

尺寸：52cm（长）x 28cm（宽）x 44cm（高）

摄影师：Sylvain Deleu. © Nuala O'Donovan

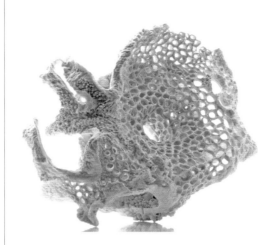

Title of work | Starfish Fractal Form

Material | High-Fired, unglazed Leach porcelain

Technique | Hand-built, Multiple firings

Dimensions | 32 x 34 x 21cm

Photographer | Sylvain Deleu. © Nuala O'Donovan.

作品名称：不规则海星型

材料：高温焙烧无光浸出瓷

工艺：手工制作，多重烧制

尺寸：32cm x 34cm x 21cm

摄影师：Sylvain Deleu. © Nuala O'Donovan

Title of work | Pinecone, Nest

Material | High-Fired, unglazed Limoges porcelain

Technique | Hand-built, Multiple firings

Dimensions | 38 x 38 x 24cm

Photographer | Sylvain Deleu. © Nuala O'Donovan

作品名称：松球巢

材料：高温焙烧里摩日素瓷

工艺：手工制作，多重烧制

尺寸：38cm x 38cm x 24cm

摄影师：Sylvain Deleu. © Nuala O'Donovan

Title of work | Coral, Dynamic Form

Material | High-Fired, unglazed porcelain

Technique | Hand-built, Multiple firings

Dimensions | 38 x 34x 30cm

Photographer | Sylvain Deleu. © Nuala O'Donovan

作品名称：动态珊瑚型

材料：高温焙烧素瓷

工艺：手工制作，多重烧制

尺寸：38cm x 34cmx 30cm

摄影师：Sylvain Deleu. © Nuala O'Donovan

Title of work | Banksia, Dynamic

Material | High-Fired, (1240 centigrade), unglazed porcelain

Technique | Hand-built, Multiple firings

Dimensions | 33 x 38 x 38cm

Photographer | Sylvain Deleu. © Nuala O'Donovan

作品名称：动态山龙眼

材料：高温焙烧（1240度）素瓷

工艺：手工制作，多重烧制

尺寸：33cm x 38cm x 38cm

摄影师：Sylvain Deleu. © Nuala O'Donovan

Title of work | Teasel Combined Patterns

Material | High-Fired, (1260 centigrade), unglazed porcelain

Technique | Hand-built, Multiple firings

Dimensions | 36 x 36 x 42cm

Photographer | Sylvain Deleu. © Nuala O'Donovan.

作品名称：起绒草组合型

材料：高温焙烧（1260度）素瓷

工艺：手工制作，多重烧制

尺寸：36cm x 36cm x 42cm

摄影师：Sylvain Deleu. © Nuala O'Donovan

WIDE ANGLE
广角

DESIGN, A LANGUAGE THAT REFLECTED THE SITE, YET IS SUPPOSED TO SYMBOLIZE OMNIPRESENT RESTRICTIONS TO CREATIVITY IN ART AND DESIGN.

VOLKSWAGEN WIRE

大众电线车型

Designer | Karen Oetling Corvera, Juan Pablo Ramos Valadez

A creation between sculpture and machine, art and design, the Volkswagen wire project was a commissioned work from a private client made by designers Karen Oetling and Juan Pablo Ramos Valadez from Guadalajara, Mexico.

It was ment to fill up a green open space with certain nostalgia with a sense of simplicity and openness, and a language that reflected the site; the strong shape would also have to merge with the green area to be environmentally unperceptive.

The scale is 1:1 to a 1989 Volkswagen beetle chassis, and the wire mesh opens up letting not only line cut it through, but creating at times an unperceptive form.

The wire framework turned into a sculpture made out of iron. The skeleton was painted green in order to blend in with the environment.

The form is true to the chassis and even opens up as the boot and doors would on the original machine.

Finally, front and back lights of the original car were attached to the wire framework to render a mental image of the real machine not however ever detracting from the sculptural piece.

It took about a month from pencil till delivery to complete and it weighted around 200 kilograms.

  一个介于雕塑和机械、艺术和设计之间的创作——大众电线车型项目，是一个由私人客户委托设计的作品，设计师是来自墨西哥瓜达拉哈拉的 Karen Oetling 和 Juan Pablo Ramos Valadez。

  这个车型被用做填补一个绿色开放空间，这个空间带着一丝特定的简洁、开阔的怀旧感。它作为一种语言用以述说这一空间。它存在感极强的外在造型也同样需要与周围的绿色区域相融合，消除环境突兀感。

  这个项目按 1:1 的比例模拟 1989 年大众出产的甲壳虫车型，电线网眼的开放不仅让线条能够穿插而过，还能不时创造出一种意想不到的形式。

  这个电线框架由于其中的铁而能够成型。整体框架被涂成了绿色，与周边环境相融合。

  这个形式就像一个真的车型，甚至它的后盖和车门都如同原型车一样能够打开。

  原型车的车灯被卸下后安装在了这个模型上，以此创造一种真车的质感而不仅仅是一座雕塑。

  从绘制模型图到最后完工大约花费了一个月时间。车型重量约 200 千克。

HOMAGE TO THE LOST SPACES

致敬基督城的失落空间

Designer | Mike Hewson

Entitled "Homage to the Lost Spaces" (Government Life Building Studios) Mike Hewson's photographic installation references the artistic community of pre-earthquake Christchurch who lost studios and seeks to draw attention to and pay final respects to the beauty contributed by many of the old buildings of the cities.

Mike Hewson was amongst many artists who lost studio space, work, and their community when the Government Life Building in Cathedral Square was demolished. When taking photos of the daily life of the artists around him, Hewson never envisaged the images to take on such significance.

"People can transform otherwise sterile and abandoned spaces into something that is full of life and vibrancy," notes Hewson, "I intended this work to project the same spirit of life back into Cranmer Courts, to help people remember, before it is gone, that this building too was once full of community, fun, and family."

Hewson 从没想过这些图片能带来如此效果。

Mike Hewson 的摄影装置项目题为"致敬基督城的失落空间",是对地震前基督城的艺术社区的一种追忆。那些失去了自己工作室的艺术家们想要引起民众的注意,对这个由许许多多旧建筑汇聚而成的美献上最后的致礼。

大教堂广场的政府生活大楼倒塌使得 Mike Hewson 与许多艺术家一样失去了自己的工作空间和社区。当对自己身边的艺术家的生活情况进行拍摄时,

"人们能够将贫瘠或是废弃的空间变得生机勃勃,"Hewson 说道,"我希望通过这一项目将这种精神带进克莱默区,帮助人们回忆起这里曾经也是有着各种社区、家庭的欢乐地区。"

LANDSCAPE ABBREVIATED

景观缩放

Designer | Nova Jiang

In her artistic practice, Nova Jiang explores unexpected connections between various fields of cultural production. She creates open systems that invite the audience to play. The visitor becomes not only spectator but also collaborator, participant and performer. For this work, Jiang creates a kinetic maze consisting of modular elements with rotating planters, which form a garden that is simultaneously a machine. The planters contain live moss collected from the sides of buildings, cracks in the pavement, subway grates and other urban nooks and crannies in New York City's landscape. Full of particles of broken glass, plastic and other detritus, they form a patchwork of unintentional archaeology.

Jiang is also interested in the way that simple interventions can make the experience of space dynamic and unpredictable. The planters are controlled by a software program that continuously generates new maze patterns based on mathematical rules; they rotate to form shifting pathways that encourage visitors to change direction and viewpoints as they move through the space.

在 Nova Jiang 的艺术经历中，她不断探索着不同领域文化产物之间意想不到的联系。她设计了一个开放系统邀请观众参与游戏。游客不仅能够成为观察者，同时也是一个合作者、参与者和实施者。在这个项目中，Jiang 创造了由带有旋转花槽的模块化元素组成的一个动能迷宫，它是一个机械系统，同时也是一座花园。花槽中种有从建筑周边、地面缝隙、下水道、地铁箅子和纽约市其他角落和裂缝中取来的苔藓，填充有玻璃、塑料碎粒或是岩屑，从而形成了一种意外的拼缀。

Jiang 同样对通过简单的干预而创造出的充满活力或意料之外的空间充满兴趣。这些花槽由软件系统控制，按照数学规则能够持续不断产生新的迷宫模式。它们不断变换，形成了一种移动的通道，鼓励参与其中的游客不断改变方向和视觉角度。

PTASZARNIA

筑巢

Design Agency | Wamhouse   Designer | Karina Wiciak

"Ptaszarnia" is the first (November 2012) of twelve parts of an original collection called "XII", entirely full designed by Karina Wiciak (designer from Wamhouse). The collection "XII" will consist of 12 thematic interior designs, together with furniture and fittings, which in each part will be interconnected, not only in terms of style, but also by name. Each subsequent design will be created within one month, and the entire collection will take one year to create.

The "Ptaszyna" lamp (which in Polish means "nestling") looks luxurious at the first glance. Gold and black finishing, as well as shape resembling a reverse crown are to symbolize splendour and wealth. But do not let the appearance deceive you, as this lamp has a second nature. Dripping golden icicles have irregular and awkward shape, which makes the lamp a little scary as well.

The "Ptaszek" armchair (which in Polish means "birdie") also looks luxuriously and resembles a bird, but the surprise is the fact that the armchair walks over the floor and the walls, leaving tracks behind it.

Thanks to this, the entire enterer comes to live and assumes slightly grotesque appearance, which reflects the author's idea on architecture and design in general – which we should treat with greater sense of humour.

"筑巢"是XII系列十二个部分中的第一个（发布于2012年11月），完全由来自Wamhouse的设计师Karina Wiciak设计。XII系列包含十二个主题式室内空间及其家具和配件的设计。每个部分不仅在形式上，同时在命名上都存在着内在联系。每个后续的设计都将耗时一个月，整个系列设计将耗时一年。

灯具"筑巢"乍看非常奢华。金色搭配黑色的细节描绘，形状类似反向皇冠，象征着荣光和富裕。但是，不要被它的外表欺骗了，因为这个灯具还有一个属性。滴落的尖锥有着不规则且不协调的形态，

使得整个灯具有种令人惊慌的感觉。

扶手椅"幼鸟"同样奢华异常,并做成了鸟的形态,但是令人惊讶的是这张椅子好像走过地面和墙壁并在后面留下了自己的行走轨迹。

多亏了这种轨迹使得这种移动感生动起来,并能设定出一种怪异的外在来表达作者对建筑和设计的整体想法,但这需要我们用一种强大的幽默感去体会和对待。

UBOJNIA

屠宰场

Design Agency | Wamhouse   **Designer** | Karina Wiciak

"Ubojnia" is the second (Dcemeber 2012)of twelve parts of an original collection called "XII", entirely designed by Karina Wiciak (designer from Wamhouse).

"Ubojnia" (slaughterhouse) is not only an interior design, but also a combination of design and art.

At first glance, "Ubojnia" shows a motif of hatching, a sketch which not only the main element of decoration, but also visually changes the scale of the room. People who stay in such an interior may get an impression that they are shrunk, or at least that they are in a fabulous and unreal world.

Yet, the very name of the establishment suggests another, hidden message (in Polish "ubojnia" means "slaughterhouse"). Seemingly-paper armchairs called "Szkic" (in Polish "szkic" means "sketch") are suspended on meat hooks or tied to chains, which is supposed to symbolize omnipresent restrictions to creativity in art and design. The very word "ubojnia" may indicate killing of talent and creation, though not necessarily by third parties (as these are not shown in the images), but by the artists and designers themselves. Yet, it is but one of numerous interpretations of symbols which anyone can understand in their own way, or not interpret at all. After all, it is a commercial interior, so any possible, more or less blatant, ideology can be treated with a pinch of salt, and in the case of this specific interior, even with slightly "noir" sense of humour.

The "Ubojnia" design includes armchair "Szkic" (which in Polish means "Sketch"), a suspended stool "Szkicownik" (which in Polish means "sketchbook"), a chandelier "Papierek" (which in Polish means "piece of paper"), A sphere-shaped, suspended and floor lamp "Kula" (which in Polish means "sphere"), a smaller suspended lamp and wall lamp "Haczyk" (which in Polish means "hook").

"屠宰场"是XII系列十二个部分中的第二个（发布于2012年12月），完全由来自Wamhouse的设计师Karina Wiciak设计。

"屠宰场"不只是一个室内设计作品，它同时也是设计与艺术的结合。

乍一看，"屠宰场"展现出一种影线图形，一种通过主要装饰元素和视觉效果改变空间大小的素描图。人们在这个房间会产生一种被挤压感，或者至少觉得他们正身处一个奇妙而虚幻的世界中。

然而，这一空间的名字还暗示着另一个信息，一条隐藏的信息。纸质的扶手椅被称做"Szkic"（波兰语意思为素描），它被悬挂在肉钩上或是绑在链条上，寓意在艺术和设计中无处不在的灵感束缚。而"屠宰场"一词则表达出对天赋和创造性的的抹杀，虽然这种抹杀常常来自于艺术家或设计师本身而非第三方。然而，每个人对这种意向的理解都有自己的看法，甚至有时会完全没有想法。毕竟，这是一个商业性的室内空间，因此不可避免地具有公众性，这时的意识形态常常被持以将信将疑的态度。在这一室内空间中甚至具有些许黑色幽默感。

"屠宰场"设计包括对扶手椅"素描"、悬空高脚凳"素描本"、枝形吊灯"纸张"、球形吊灯、落地灯"球体"和一个更小的吊灯和壁灯"挂钩"的设计。

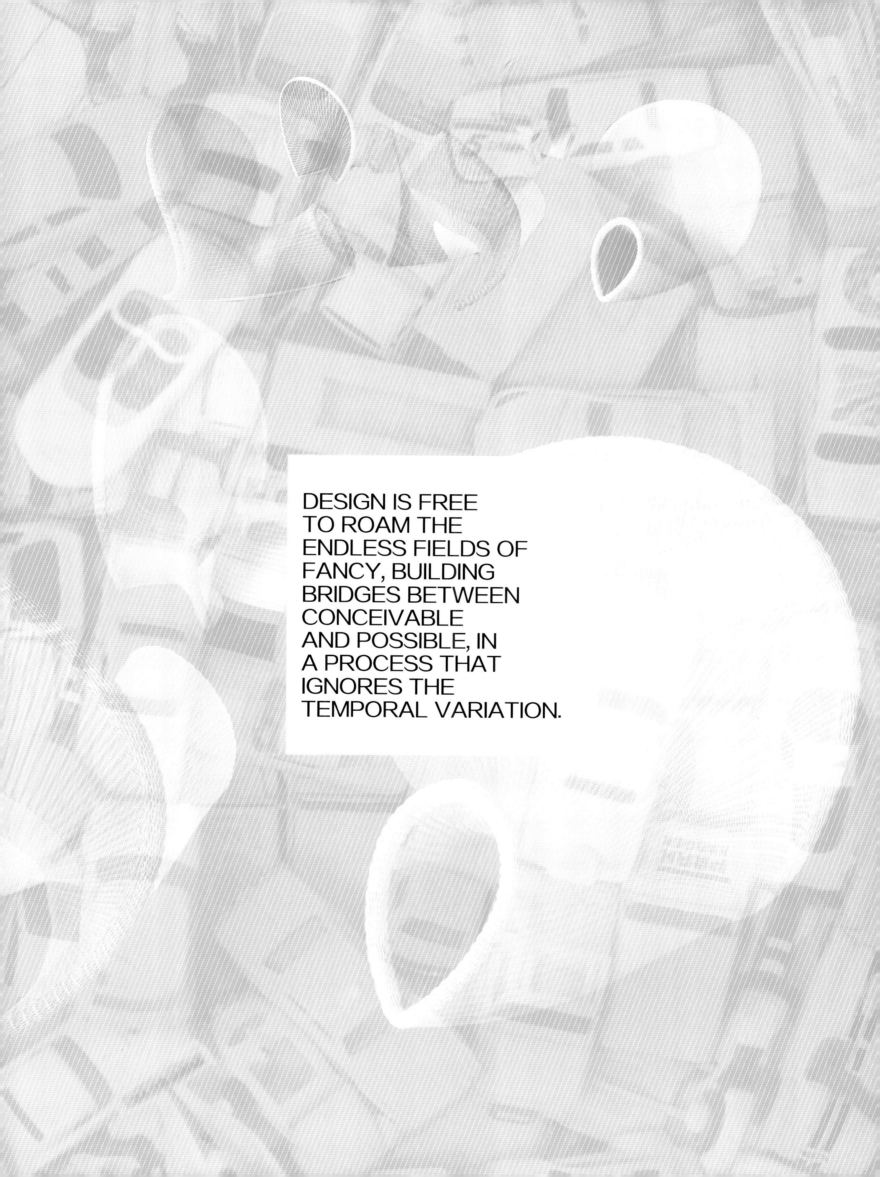

DESIGN IS FREE TO ROAM THE ENDLESS FIELDS OF FANCY, BUILDING BRIDGES BETWEEN CONCEIVABLE AND POSSIBLE, IN A PROCESS THAT IGNORES THE TEMPORAL VARIATION.

PIANTALÀ

清新篱笆

Design Agency | Andrea Rekalidis Studio

For many urban and small space gardeners a white picket fence is only but a dream. A dream we get to visit when we travel through the suburbs or small towns.

Italian designer Andrea Rekalidis has re-imagined the bucolic white picket fence as a portable and durable metal trellis.

PiantaLà can be used as a trellis where climbing vines can create a living privacy screen on a porch, deck, balcony, patio, or rooftop garden when inserted into planters.

Those gardeners are lucky enough to have land to plant in – that want the look of a picket fence without the labor – can sink it into the ground.

对许多城市和小空间园艺师而言，白色尖桩篱笆就像一个梦，一个当我们穿越郊区或小镇时才能享有的美梦。

意大利设计师 Andrea Rekalidis 重新构思了田园风格的白色尖桩篱笆，将它改造成便携式的金属交织格状框架。

这款清新篱笆能够当做藤蔓植物的支架，在门廊、露天平台、阳台、露台或是屋顶花园中形成保护生活隐私的一面屏风。

那些有幸拥有土地的园艺师们可以直接将它插进土里，这样便能在他们的土地上不费力气地打造出尖桩篱笆的效果。

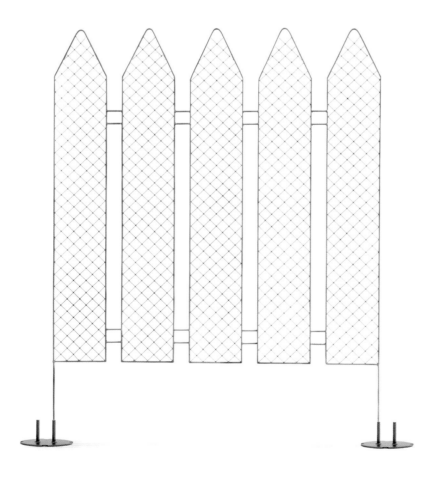

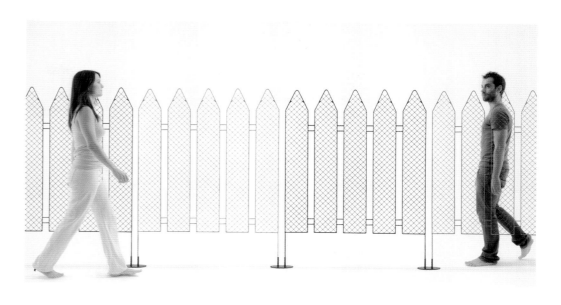

IDEAS | ARTPOWER CREATIVE SPACE

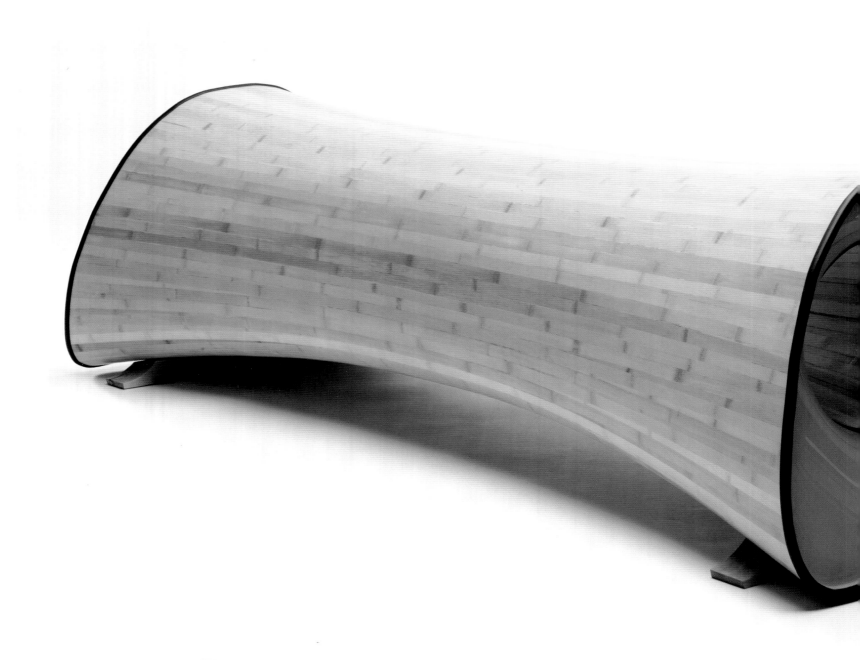

INFINITY BENCH

"无限"长凳

Designers | Andrew Williams, Tom Huang

"Infinity" is a bench that explores the possibilities of bamboo strip construction. The process of designing and constructing the bench was done by Andrew Williams and Tom Huang. The idea came after a set of three cocktail tables built using the same techniques. The designer wanted to see if the strip making process, originally used in the construction of wooden canoes, was structural enough to be implemented into a seating project. The "infinity" bench is designed as public seating for the lobby of a museum or gallery.

"无限"是一张探寻竹条构建的各种可能性的长凳。长凳的设计和制造过程由 Andrew Williams 和 Tom Huang 执行。设计灵感来自于一系列使用同种工艺制作的三张咖啡茶几。设计师们希望看看这种原本应用在木船制造过程的条状工艺是否足以应用在座椅项目中。"无限"长凳被设计作为博物馆或画廊大厅的休息座椅。

NOTHING IS QUITE
AS IT SEEMS

事事非所见

Designer | Arik Levy **Photographer** | Ian Scigliuzzi

"Nothing is quite as it seems", a retrospective and prospective exhibition by Arik Levy, to retrace his creative path ever since his very first exhibition in 1986.

"Nothing is quite as it seems" because Arik Levy investigates the relationship between visible and invisible, full and empty, existing and absent, tangible and intangible.

Based on these assessments, Arik Levy works on the evolution of the Rock, one of his major pieces. This evolution relates the absence of parts that were taken off as well as the light: the reflection of images and light coming from other angles, invisible to our eye when we look at a fixed direction. Every facet of these elements represents the absence of nature as it is not made nor really looks like real nature, but rather like a nature from a different civilization.

"事事非所见"是 Arik Levy 的回顾—展望系列展览。这一展览追溯了 Arik 自 1986 年首展之后的创作历程。

"事事非所见"得结论于 Arik Levy 对可见与不可见、满与空、存在与消失、有形与无形之间关系的研究。

基于这一研究，Arik Levy 致力于石头的演化，这个石头也成为他的主要作品之一。这种演变与去除石头的某些部分以及光线的变化有关：一些画面的反射和来自其他角度的光线，当我们的视线固定于某个方向时并不能为我们所见。这些几何体的每一面都代表着自然的暂时缺席，因为它并不是由自然组成，看着也并非真正的自然，反倒像来自不同文明的自然。

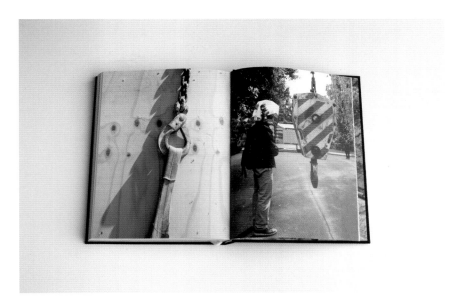

THE CANDELIER

小熊吊灯

Designer | Kevin Champeny

There are two sizes of the "Candelier". The largest one is 78.74cm diameter and the smaller one shown in the photos is 45.72cm diameter. Both are built with about 30.48cm of space in between the ceiling plate and the top of the sphere of bears. There are about 3,000 bears in the 45.72cm diameter version and over 8,500 in the 78.74cm diameter version. All of the bears are hand cast acrylic. The designer sculpted the original gummibear and produced a silicone rubber mold and hand cast all of the gummibears himself in his Westchester studio. He handed string each bear on the line with a small knot and a tiny glass bead to hold it in place. The smaller version takes about 3 weeks to produce, while the larger version can take up to several months.

这款小熊吊灯有两种尺寸：直径 78.74 厘米及直径 45.72 厘米（图片为 45.72 厘米）。灯顶与天花板之间相距 30.48 厘米。45.72 厘米的吊灯挂有约 3000 只小熊而 78.74 厘米的小熊超过 8500 只。所有的小熊都是手工切割亚克力制成。设计师刻出了小熊的最初形态并制作了硅橡胶模具，并在他的工作室中亲手进行浇铸。他为每只小熊穿上线，打上绳结，并用一个小玻璃珠将它固定。小的吊灯制作耗时约 3 周，而大的需要几个月时间。

IDEAS | ARTPOWER CREATIVE SPACE

THE HOTWHEELS WALL ART

"赛车"墙面艺术

Designer | Kevin Champeny

Kevin Champeny created the "Hotwheels" Logo (Hotwheels is a registered trademark of Mattel) out of over 4,400 hotwheels. It weighs over 249.48kg, is 2.74m long x 1.22m tall. It was a custom project he built for a car collector in Canada. It took him over 6 months to create. Each car is left in its original form, he did not paint any of them. He hand glued each car with almost 3 layers of cars to create the overall effect.

Kevin Champeny 用 4400 辆玩具赛车制作出了"风火轮"logo（风火轮是美泰公司的注册商标）。它们超过 249.48 公斤，2.74 米长、1.22 米高。这是加拿大一个汽车收集者的客户定制项目，设计师花费了超过 6 个月的时间进行创作。每辆车都以最原始的状态呈现，没有经过任何上色。设计师用胶水黏合起约 3 层的车来制作出想要的整体效果。

ARTPOWER CREATIVE SPACE | IDEAS

THE ICE LAMP

冰灯

Designer | Kevin Champeny

The Ice Lamp is 30.48cm diameter and contains over 500 pieces of hand cast rubber shards of glass and sells for $350.00 US. Kevin Champeny created this lamp out of a desire to experiment with perception. He wanted to create an object that was harsh, jagged, and sharp to the eye, but in reality was soft enough to rest your hand on. The juxtaposition between what it looks like and what it actually is creates a playful tension for the viewer. He broke several wine bottles and created molds made from Silicone and then hand dye small batches of rubber and cast them into the molds. Each lamp is hand made and meticulously fabricated to give the appearance of a ball of sharp glass. He cast 8 shades of blue so when the lamp is off, it still has the feeling of being on. The light blue hue of the lamp when it is turned on is also very serene.

　　这个冰灯直径30.48厘米，包含超过500片手工浇铸的仿玻璃碎片橡胶，售价350美元。Kevin Champeny 创造这个灯具是希望对知觉进行实验。设计师希望创造出一个物体，粗糙、交错不齐、刺人眼球但实际上却很柔软可触。视觉上和实际上的矛盾使观看者产生了一种紧张感。他打碎了几个酒瓶，用硅做出了模型，接着将小批量塑胶进行手工染色并放入了模型中。每个灯都是手工制作，细致地描绘出一个球体的尖锐玻璃外表。他涂了八个深度的蓝色，当灯灭时也像依然亮着。淡蓝色的色度在灯打开时依然显得非常柔和。

PENDANT CHANDELIER

垂饰吊灯

Designer | Naomi Paul

In the OMI pendant collection, GLÜCK, SONNE and HANNA are the principal designs in a series of elegant flat pack pendants, which evolved from the desire to create timeless, playful and quietly opulent lighting solutions. Whether lit or not the clean finely crafted silhouettes form a modern yet sculptural focal point within any space. This season welcomes new pendant MONIKA, a large-scale pendant chandelier with multi-bulb fittings along side XL size editions for all the shapes in the collection.

The pendants are constructed by hand in Great Britain using traditional crochet techniques, made with yarns such as mercerized cotton and silk. Where normally this pre-consumer yarn waste would be discarded, here it is utilised to create stunning lighting products. All parts and materials are sourced as locally as possible.

在OMI垂饰系列中，GLÜCK，SONNE和HANNA是一系列典雅的扁平封装垂饰中的主要设计，源于对创造永恒的、可玩的、低调华丽的照明方式的渴望。无论是否点亮，这些造型简洁细致的剪影在任何空间中都能形成一种现代而具有雕塑般聚焦性的效果。同这一系列的其他XL型号和各色形态的产品一起，MONIKA作为有着多个灯泡的大规格吊灯，成为这一季度非常值得期待的新品。

这些垂饰来自于英国手工钩针编织工艺，由纱线制成，包括丝光棉、丝绸。这些平常使用过的丝线本要被丢弃，但这里它们被再利用后成为了美丽的照明产品。所有材料都尽可能取自于当地。

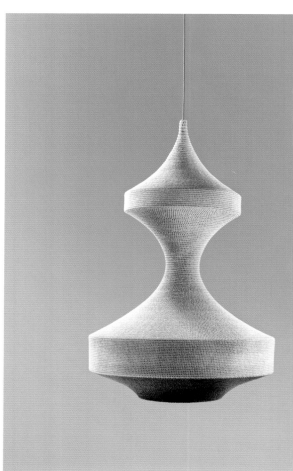

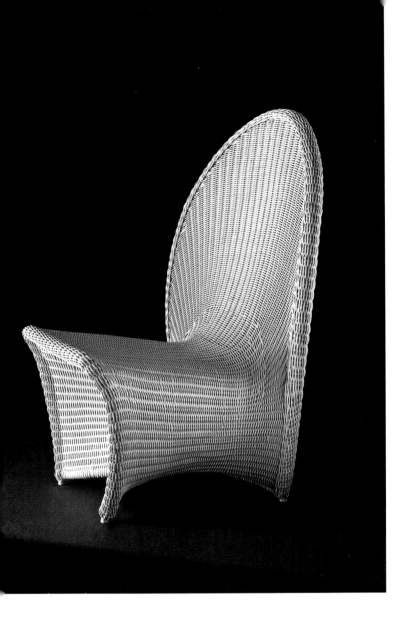

36H AND 56H

36 小时 /56 小时

Designer | Fabio Novembre

One of Fabio Novembre's earliest memories dates back to when his grandmother prepared the homemade pasta. He was ecstatic by the ritual she staged: the wisdom of the dough, the homogeneous leveled mass, the domestication of the surfaces.

Gestures handed down for generations.

The world has changed so much since then, and practice and imagination refer less and less to a single actor.

The designer is free to roam the endless fields of fancy, building precarious bridges between conceivable and possible, in a process that ignores the temporal variation. On the contrary, those who turn his thoughts into substance, through manual dexterity and mastery of gestures, consider time as a plus value. The names of these two objects of memory, fellow adventurers in one's raids of fantasy, refer then to the temporal element that each of them requires ,as a pledge in the hands of skilled craftsmen, in honor of their work and of the value that they'll assume for those who will own them.

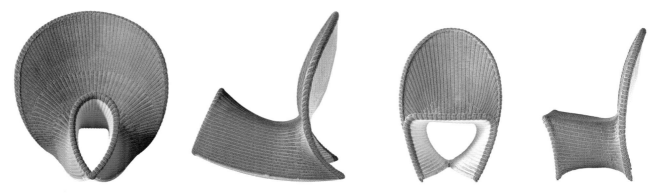

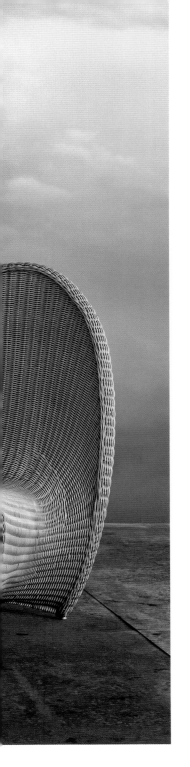

Fabio Novembre 最早的记忆可以追溯到奶奶为其准备自制面条。他对面条的制作过程心醉神迷：发面团，和面，擀面。

这些姿势一代代传下来。

这个世界自那时起已发生了巨大改变，一些行为和预想越来越不参照于单一的展示者。

设计师遨游于无边的幻想之中，忽略时间的变化，在想象和可能之间搭建出不稳定的桥梁。与此相反，那些能够将想法通过灵巧的手工和精湛的技艺变成实体的人，则认为时间是一个附加值。这两个作品，突现于某人刹那的幻想中，它有关记忆与一同冒险的伙伴，其命名，与时间元素相关。它们成为了技工师手中的抵押品，作为对他们的作品以及这些作品对拥有他们的人形成的价值的一种纪念。

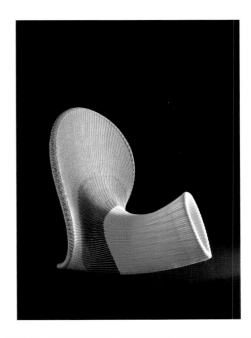

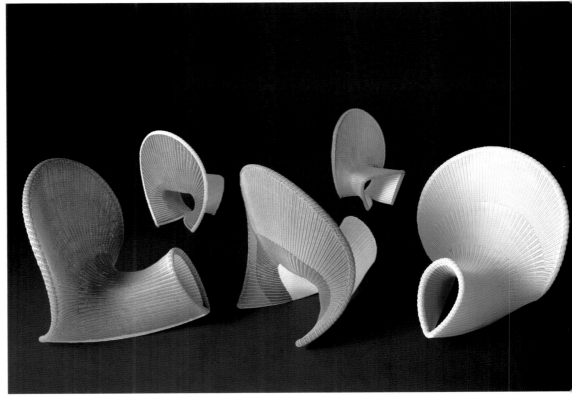

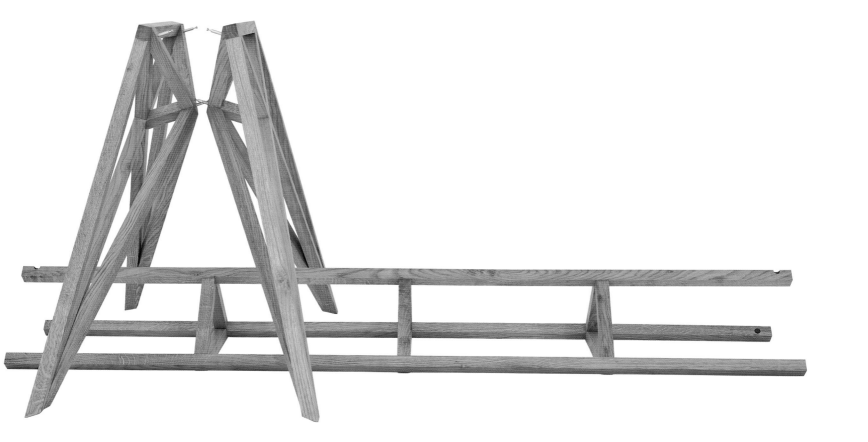

PONTE

桥式桌

Designer | Tom Strala

A few years ago Tom Strala stired up the design community. His TMS lights unite consequent architecture with artful sensuality. In the same spirit his recent object presents itself: a table, eluding fashion and movements, an everyday object with that special something. "The table is constructed from a 22.5m long wooden piece which in itself is no quality. Still this clarity has something beautiful, since the sententiousness of the final shape includes the simplicity, and does not negate it." says Tom Strala about the object Ponte. The base frame consists of massive oak wood elements, a composition in framework style. It is mounted with only 6 screws. Its table top is made of clear glass.

几年前 Tom Strala 在设计界激起了波浪。他的 TMS 灯具将建筑与艺术感受进行结合。本着同样的精神他最近的作品很好地展示了这一点：一张桌子，剔除了时尚元素和活动性，作为一个日常用具却拥有一些独特特质。"这张桌子由一根长 22.5 米的木头制成，木头本身并不能显示出品质，但它以最后的简洁形态展示出一种简单的美丽，这一点我们无法忽略。" Tom Strala 说到。整个基础结构包括大量的橡木元素，形成一种组合式的框架风格。它仅由 6 根螺丝接连，桌面由透明玻璃制成。

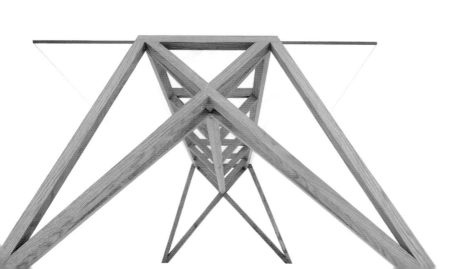

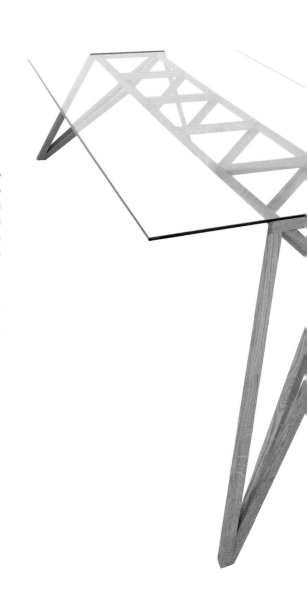

"XII" COLLECTION – MASARNIA

"XII" 系列——MASARNIA

Designer | Karina Wiciak

"Masarnia" is the January (2013), and at the same time the third, design from the "XII" collection, entirely designed by Karina Wiciak.

At first glance, "Masarnia" looks like a sweet, winter-Christmas picture, or like a modern, elegant interior. Glass furniture and lamps give the impression of wealth, and the classic, and red color is a decorative feature in the minimalistic, white interior.

Yet, on closer inspection it turns out that the interior is more macabre than sweet.

It turns out that the white furniture and lamps are rescaled elements of the human body, and the red elements of the design imitate blood. As you can see, the glass decorations with sharp edges are also not random, as, according to the author's conception, they were used to make these gruesome furniture.

The design symbolically shows objective treatment of the human body, e.g. when creating the so-called consumer goods (including design).

The author believes that if people can destroy or exploit the nature, then perhaps it is worth to think what would happen in the roles were reversed? Yet, this design is neither an ecological manifesto, nor an attempt to preach. "Masarnia" (like the previous design "Slaughterhouse") is merely a pretext for reflection and another concept of an interior with a slight (perhaps black) sense of humour.

  Masarnia 是"XII"系列 2013 年一月的作品，同时也是这个系列的第三个作品，它完全由 Karina Wiciak 设计。

  乍一看，Masarnia 就像一个甜蜜的冬季圣诞图片，或是一个现代、优雅的室内展示。玻璃家具和灯具产生一种富裕的印象，古典的红色是这一极简白色室内设计的装饰特色。

  近距离观察，这一室内空间与其说是甜蜜的，不如说是骇人的。

  这些白色家具和灯具都是人体部位的缩放，设计中的红色意味着血液。正如所见，有着尖锐边缘的玻璃装饰并不是随机的，而是按照作者的设想，制造出阴森的家具质感。

  这个设计象征性地展现出对人体的客观对待，例如，创造出所谓的消费品（包括设计）。

  作者相信如果人们能够破坏或利用自然，那么当情况相反时会发生什么呢？然而这个设计并不是一个生态宣言也不是一次宣讲的尝试。Masarnia 仅仅是一个反思的载体，以及对室内设计的另一个带有些许黑色幽默的概念。

WATERTOWER

水塔

Designer | Tom Fruin Studio   **Dimension** | 609.6 x 304.8 x 304.8cm   **Photographer** | Robert Banat

Tom Fruin Studio is pleased to present Watertower, a new sculptural artwork by Brooklyn artist Tom Fruin. For the US premiere of his internationally recognized Icon series, Fruin has created a monumental water tower sculpture in colorful salvaged plexiglas and steel. Watertower is mounted high upon a water tower platform becoming part of the DUMBO, Brooklyn skyline.

Fruin, who often works with reclaimed and discarded materials, has composed Watertower from roughly one thousand scraps of plexiglas. The locally-sourced plexi came from all over New York City.

Illuminated by the sun during the day and by Ardunio-controlled light sequences designed by Ryan Holsopple at night, this beacon of light is a tribute to the iconic New York water tower and a symbol of the vibrancy of Brooklyn.

Tom Fruin Studio 很高兴能够带来由布鲁克林艺术家 Tom Fruin 设计的新雕塑艺术品——水塔。继在美国首次亮相的、获得国际认可的图标系列之后，Fruin 创作出了一个具有纪念意义的水塔。这个水塔由色彩鲜艳的废弃树脂玻璃和钢材制成。水塔矗立在真正的水塔平台之上，成为布鲁克林天际的一部分。

Fruin 常常回收利用废弃材料，水塔就是由约 1000 个树脂玻璃碎片组成。这些玻璃来自于全纽约市。

水塔在白天由阳光点亮，夜晚由 Ryan Holsopple 设计的阿依诺系统控制灯进行照明。这座水塔是纽约标志性水塔的又一力作，也是充满活力的城市布鲁克林的象征。

LEVITY
CHAISE-LOUNGE

飘浮轻便躺椅

Designer | Zsanett Benedek & Daniel Lakos   Photographer | Tamás Bujnovszky

The Levity chaise-lounge is composed of a 12 mm thick birch plywood and 12 mm thick birch rods. Although the plywood shell is quite thin, there are twenty legs supporting it. The wooden rods are attached firmly in the drilled holes. This simple construction is lightweight and surprisingly sturdy.

这款飘浮轻便躺椅由一片 12 毫米厚的桦木胶合板和多条 12 毫米粗的桦木棒条组成。虽然胶合面板十分轻薄,但 20 条支架足以支撑。这些木质支架被牢牢地扣在了钻孔中。这个简单的结构轻便但异常坚固。

名門八十八號
NO.EIGHTY-EIGHT MANSION

Luxury Penthouses Continental Manors
—— Manor 88 in Xianyou

空中别墅 欧陆庄园
——仙游名门八十八号

The top priority of high-end properties is about the experience of living quarter, such as the value of rare location, public resources and landscaping being the key factors of buyers choosing their houses, especially under the circumstance that urban space tends to be highly intensified in modern time. Manor 88 incorporates all the above-mentioned value factors together by integrating magnificent continental manors, over 600 ingenious penthouses and French-style royal garden planning, which features inventive design over the slopes and flat ground. Residents strolling within the prestigious community will certainly appreciate every different view with every step.

高端楼盘首先注重的是人居体验，在城市的土地空间高度集约化的情况下，稀缺的区位价值、公共资源的价值、景观价值已成为置业选择的重要因素。名门八十八号整合这三类价值于一体，不仅规划有奢华的欧陆庄园别墅，又有600余套独具创意的空中别墅，更通过坡地的合理利用，营造出立体化多层次的法式皇家园林景观，让生活在城市的人们也能享受到"庭院深深深几许，步移景异景时移"的别样意境。